# 9th International Conference on Advanced Fibers and Polymer Materials (ICAFPM 2019)



Shanghai, China
19 - 22 November 2019

ISBN: 978-1-7138-0622-6

**Printed from e-media with permission by:**

Curran Associates, Inc.
57 Morehouse Lane
Red Hook, NY 12571

**Some format issues inherent in the e-media version may also appear in this print version.**



Printed with permission by Curran Associates, Inc. (2022)

For permission requests, please contact Donghua University
at the address below.

Donghua University – College of Materials Science & Engineering
2999 North Renmin Road
Songjiang, Shanghai
China, 201600

Phone: 0086-21-67792362
Fax: 0086-21-67792855

http://cmse.dhu.edu.c

**Additional copies of this publication are available from:**

Curran Associates, Inc.
57 Morehouse Lane
Red Hook, NY 12571 USA
Phone: 845-758-0400
Fax: 845-758-2633
Email: curran@proceedings.com
Web: www.proceedings.com

ICAFPM

# ICAFPM2019

## 第九届先进纤维与聚合物材料国际会议

THE 9TH INTERNATIONAL CONFERENCE ON ADVANCED FIBERS AND POLYMER MATERIALS

# ABSTRACTS

2019年11月19-22日 中国·上海

*Nov. 19th-22nd, 2019 Shanghai, China*

SKLFPM
DONGHUA UNIVERSITY
纤维材料改性国家重点实验室

纤维材料改性国家重点实验室(东华大学)
State Key Laboratory for Modification of Chemical Fibers and Polymer Materials (SKLFPM, DHU)

## 主办(ORGANIZER)

纤维材料改性国家重点实验室(东华大学)
State Key Laboratory for Modification of Chemical Fibers and Polymer Materials (SKLFPM, DHU)

SKLFPM
DONGHUA UNIVERSITY
纤维材料改性国家重点实验室

## 承办(CO-ORGANIZER)

中国材料研究学会纤维材料改性与复合技术分会
Division of Fiber Materials and Composite Technology, Chinese Materials Research Society (FMCT, CMRS)

纤维材料先进制造技术与科学创新引智基地
Innovation Base of Advanced Fabrication Technology of Fiber Materials, DHU (111 Project)

东华大学材料科学与工程学院
College of Materials Science and Engineering (CMSE, DHU)

111 PROJECT
111 引智基地

## 协办(SPONSOR)

教育部材料类专业教学指导委员会高分子材料工作组
Polymer Materials Working Group in Material Division of Teaching Steering Committee of Ministry of Education

高性能纤维及制品教育部重点实验室(B)
The Key Laboratory of High-Performance Fibers and Product, Ministry of Education

东华大学先进低维材料中心
Center for Advanced Low-dimension Materials (CALM, DHU)

上海市先进纤维与低维材料一带一路国际联合实验室(东华大学)
Shanghai Belt and Road Joint Laboratory of Advanced Fiber and Low-dimension Materials (Donghua University)

上海市轻质结构复合材料重点实验室
Shanghai Key Laboratory of Light Weight Structural Composite Materials

功能杂化材料上海高校重点实验室
Key Laboratory of Hybrid Functional Materials of the Universities in Shanghai

# ICAFPM 2019

# INDEX

## Abstracts of Plenary Session ........ P1-P20

# INDEX

## Abstracts of Session A ........................ A1-A29

# INDEX

# INDEX

# INDEX

# INDEX

# INDEX

# INDEX

# INDEX

# INDEX

# INDEX

# INDEX

# INDEX

# INDEX

# INDEX

# INDEX

# INDEX

# INDEX

# INDEX

# INDEX

# INDEX

# INDEX

# INDEX

# INDEX

# INDEX

# INDEX

# INDEX

# INDEX

# INDEX

# INDEX

# INDEX

**Dr., Prof. Baughman, Ray H.**
**University of Texas at Dallas**

Ray Baughman received a B.S. in Physics from Carnegie Mellon University and a Ph.D. in the Materials Science area from Harvard University. Upon graduation he went to Allied Chemical, which later became AlliedSignal and then Honeywell. In August 2001, he became the Robert A. Welch Chair in Chemistry and Director of the Alan G. MacDiarmid NanoTech Institute at the University of Texas in Dallas.

Ray is a member of the National Academy of Engineering, the Academy of Medicine, Engineering and Science of Texas, the Academia Europaea, and the European Academy of Sciences and Arts; a foreign member of the European Academy of Sciences; a Fellow of the Royal Society of Chemistry, the National Academy of Inventors, and the American Physical Society; an Academician of The Russian Academy of Natural Sciences; an honorary professor of 8 universities in China; and is on editorial or advisory boards of Science and other journals.

Ray has 90 issued US patents and over 440 refereed publications, with over 40, 500 citations and a Web of Science h-index of 88. He has received many awards including the Chemical Pioneer Award of the American Institute of Chemists (1995), the Cooperative Research Award in Polymer Science and Engineering (1996), the New Materials Innovation Prize of the Avantex International Forum for Innovative Textiles (2005), Nano 50 Awards from Nanotech Briefs Magazine for Carbon Nanotube Sheets and Yarns (2006) and for Fuel Powered Artificial Muscles (2007), the NanoVic Prize from Australia (2006), the Scientific American Magazine 50 recognition for outstanding technological leadership (2006), the CSIRO Metal for Research Achievement (2006), the Chancellor's Entrepreneurship and Invention Award (2007), the 21 for the 21st Century award (2007), the Alumni Distinguished Achievement Award of Carnegie Mellon University (2007), the Kapitza Metal of the Russian Academy of Natural Sciences (2007), the Graffin Lectureship of the American Carbon Society (2010), the Tech Titans Award in Education (2011), Time Magazine recognition in 50 Best Inventions of the Year (2011), and many others...

# Sheath-Run Artificial Muscles and Their Use for Robotics, Environmental Energy Harvesters, Comfort Adjusting Textiles, and Electricity Generation

Baughman, Ray H.
University of Texas at Dallas

Remarkable performance has been obtained for tensile and torsional carbon nanotube hybrid yarn muscles, whose actuation is driven by the volume change of a guest within a twisted or coiled carbon nanotube yarn. During thermally-powered contraction, coiled hybrid muscles can deliver 29 times the work as the same weight human muscle. We here describe our new sheath-run artificial muscles (SRAMs), which eliminate major problems that exist for our carbon nanotube hybrid muscles: (a) carbon nanotube yarns are expensive and (b) yarn guest near yarn center does not efficiently contribute to actuation. An actuated guest sheath on a twisted or coiled polymer fiber or yarn drives actuation for our new sheath-run artificial muscles. This configuration change dramatically increases muscle power and enables cheap commercialized yarns to replace expensive carbon nanotube yarns. As electrochemical SRAM generates 1.98 W/g of average contractile power - 40 times that for human muscle and 9.0 times that of the highest power alternative electrochemical muscle. Electrothermal PEO-SO3@CNT SRAMs operated in air and in room-temperature water to produce 2.6 W/g (at 9 Hz) and 9.0 W/g (at 12 Hz) of full-cycle contractile power, respectively, compared with the 0.05 W/g typical of human muscle. SRAMs are woven to make textiles that automatically change porosity in response to perspiration and temperature for comfort adjusting clothing. In combination, with our twistron yarn mechanical-to-electrical energy converters, the SRAMs can be used to harvest waste chemical or thermal energy as electrical energy.

**Dr., Prof. Faul, Charl FJ**
**University of Bristol**

Charl FJ Faul received his M.Sc. (Physical Chemistry) from the Department of Chemistry, University of Stellenbosch, South Africa, investigating the destruction of halogenated wastes by means of high-temperature pyrolysis. He then spent one year in the city of Durban, on the east coast of South Africa, in the Department of Chemical Engineering, University of Natal, investigating the use of ultrasound for water treatment (a joint project with ESKOM, the national electricity provider). He then returned to Stellenbosch and completed his Ph.D. (Polymer Science) with Prof. Ron Sanderson in 2000.

Charl was visiting professor in the Molecular Materials group at the Molecular Materials Group, Helsinki University of Technology, hosted by Prof. Dr. Olli Ikkala (2006 – 2010). He then held a prestigious visiting professorship from the Chinese Academy of Sciences (hosted by Prof. Zhixiang Wei, National Centre for Nanoscience and Technology, Beijing, China) in 2012. He was furthermore invited by Prof Xi Zhang (Department of Chemistry, Tsinghua University, Beijing, China) as visiting international lecturer in 2011 and 2012, and visits Tsinghua on a regular basis. Charl was appointed in October 2013 as adjunct professor at the Department of Chemistry at Tsinghua University. This is a 3-year appointment, which has been extended for a further 3-year period. He was also visiting professor at the University of Rennes 1, Rennes, France in 2018.

He was appointed to the International Editorial Board of the newly launched Wiley journal, ChemistrySelect, in December 2015.

He is, since August 2018, the head of Inorganic and Materials Chemistry and forms part of the senior management team of the School of Chemistry.

# Electroactive Polymeric Materials – from Supramolecular Polymers to 3D Networks

Faul, Charl FJ
University of Bristol

In the quest for new materials to help to solve global challenges, we have focused our efforts in recent years on the production of novel functional polymeric materials – both supramolecular polymers and highly cross-linked covalent 3D polymer architectures. We are therefore designing new functional materials in key areas for further application:
Supramolecular Polymers: We recently explored the use of crystallisation-driven self-assembly for the formation of functional 1D micellar nanowires.2 We are now extending this approach and are preparing a platform of functional small aniline3 and perylene-based4 molecules for the production of addressable and tuneable supramolecular polymers.
3D Materials and Devices: Using related synthetic approach, we also show how aniline-based materials can be used for the production of redox-active conjugated microporous polymers (CMPs). We not only have the ability to carefully tune porosity and pore size distributions, 5a but also tune CO2 uptake with high selectivity.5b Initial results of the application of these materials for energy storage and conversion will be shown.6

References:
[1] C. F. J. Faul, Acc. Chem. Res., 2014, 47, 3428; Y. Chang, et al., Chem. Soc. Rev. 2019, 48, 989
[2] X. Li, P. Wolanin et al., Nat. Comm., 2017, 8, 159090; C. Jarret-Wilkins et al., Macromolecules 2019, 52, just accepted
[3] A. Bell et al., J. Am. Chem. Soc., 2015, 137, 14288; W. Lyu et al., Chem. Sci., 2018, 9, 4392
[4] C. Jarret-Wilkins et al., Chem. Eur. J., 2018, 24, 15556; G. Echue et al., Chem. Eur. J., 2015, 21, 5118
[5] J. Chen et al., Angew. Chem. Int. Ed. 2019, 58, 11715; H. Wang et al., Chem. Mater., 2017, 29, 4885
[6] Y. Liao et al., ACS Appl. M

**Dr., Prof. Reneker, Darrell**
**The University of Akron**

Darrell H. Reneker, Distinguished Professor of Polymer Science, has retired after 30 years of service to The University of Akron (UA). He began working on polymer physics at the DuPont Company in 1959, after receiving his Ph.D. in Solid State Physics at the University of Chicago. In 1969, he joined the National Institutes of Standards and Technology (NIST) and became manager of the Center for Materials Science. He also served as executive secretary of the Committee on Materials of the White House Science Office.

Reneker came to UA as a professor of polymer science and director of the Institute of Polymer Science in 1989.

He is recognized internationally as a leader in polymer nanofiber electrospinning — a process that creates an electrified fluid jet that coils and eventually solidifies as a continuous thin fiber. This technique has been widely used by scientists and engineers to create innovations ranging from bandages that release medication to improved filtration devices, biomedical implants and scaffolds for cell growth.

Reneker was a recipient of the Silver Medal Award from the Department of Commerce and appointed as a senior executive of the U.S. Civil Service. In 2013, Reneker was elected as a Fellow of the National Academy of Inventors. Academic inventors elected to Fellow status have demonstrated a prolific spirit of innovation in creating or facilitating inventions that have made a tangible impact on quality of life, economic development, and the welfare of society.

# Inside Nanofibers toward NANOWARE Devices

Reneker, Darrell
The University of Akron

Our work is aimed at NANOWARE devices, made, for example of polymer or carbon nanofibers, to communicate between biological molecules, prosthetics, and computers.

Segmental motions of PVDF molecules, were observed in a series of very high resolution electron micrographs of thin PVDF nanofibers in our recent work.

Excellent control of nanofiber diameter was achieved by computer control of the shape of the droplet from which the electrospinning jet flowed.

Our continuing efforts are to use conducting nanofibers to influence the functions of nerves in living organisms.

**Dr., Prof. Kitagawa Hiroshi**
**The University of Kyoto**

Dr. Kitagawa is interested in cooperative quantum properties that protons and electrons show in solid materials, such as ferromagnetism and superconductivity. Recently, he is also interested in investigating hydrogen storage, super-protonic conductivity and the photolysis of water. This focus covers a wide range of materials, from organic to inorganic matter, such as transition metal complexes, metal nanoparticles, coordination polymers and mixed-valence compounds.

He has reached more than 13, 000 citations and have won many awards including The Chemical Society of Japan Award for Creative Work (2010); Inoue Prize for Science (2011); Marco Polo della Scienza Italiana (2013).; European Advanced Materials Award (2014); The Commendation for Science and Technology by the Minister of Education, Science & Technology (2016); Fellow of the Royal Society of Chemistry (FRSC) (2016); DFG Mercator Fellow (2018); and he is appoeined as Honorary Professor at Xi'an Jiaotong University in 2019.

# Solid-State Protonic in Coordination Polymers

Kitagawa Hiroshi
The University of Kyoto

Solid-state protonics is a new research field attracting much current attention. One of the most urgent subjects in this field is to create a novel proton conductor, from the viewpoint of developing new energy and energy conservation technologies, including photovoltaic, hydrogen storage and fuel cell technologies. In this work, we report on a proton-conductive organic-inorganic hybrid system, which is a coordination polymer. Such a metal-dimer system with multi-redox property has a large potentiality for the creation of new-functional and high-performance materials in metal-complex solids.

We have developed several kinds of proton conductors, which are 0-D, 1-D, 2-D and 3-D coordination polymers. From the complex-plane impedance measurements, all the coordination polymers were found to be highly proton-conductive at room temperature. Among them, Cu rubeanate, H2dtoaCu, exhibits highest proton conduction (~ 10-2 S cm-1). This value is comparable to that of Nafion, which is famous for a solid electrolyte of fuel cell. The mechanism of proton conduction is discussed in detail. New highly proton-conductive coordination materials, conductive mixed-valent nanotube-MOF and highly-concentrated hydrogen-storage nano-materials are also presented. Structural transformation of water clusters confined in nanospace is discussed in details.

**Dr., Prof. Yu, Jianyong**
**Donghua University**

Dr. Jianyong Yu, Academician of the Chinese Academy Engineering, professor of Textile Science & Engineering, doctoral advisor, responsible for National Key Discipline of Textile Science & Engineering, is engaged in research on the basic theory, key technology and application of textile materials.

Dr. Jianyong Yu is currently appointed as the President of Donghua University. Dr. Jianyong Yu has authored and co-authored several national and international publications and also working as a reviewer for reputed professional journals. Dr. Jianyong Yu is having an active association with different societies and academies around the world. Dr. Jianyong Yu made his mark in the scientific community with the contributions and widely recognition from honourable subject experts around the world. Dr. Jianyong Yu has received several awards for the contributions to the scientific community: The Key Technology of Spinning, Weaving, Dying and Finishing of Soybean Protein Composite Fiber and the Development of Its Textile Products won National Second Grade Award for Science and Technology Progress in 2007; The Key Technology and Application of Compact Spinning System won National Second Grade Award for Technological Invention in 2007; The Composite Spinning Machine, Technology and Products of Filament and Staple Fiber won Shanghai Municipality Second Grade Award for Technological Invention in 2006; The Key technology and Application of High-quality and Low-consumption Processing of Textiles won China National Textile and Apparel Council Second Grade Award for Science and Technology Progress in 2006; The New Application Technology of Chemical Fiber and Development of Its Products Blending with Wool won Shanghai Municipality Second Grade Award for Science and Technology Progress in 2005.; esearch on the Characteristics and Processing Technology of Colored cotton Fiber obtained Shanghai Municipality Third Grade Award for Science and Technology Progress in 2004; Composite Spinning of Filament and Staple Fiber and Method of Systemic Compensation won Ministry of Education Second Grade Award for Science and Technology Progress in 2004.

# Functional Nanofibrous Materials

Yu, Jianyong
Donghua University

Nanofibers, as the forefront of advanced fibrous materials, hold great promise for improving the performance and extending capabilities of products in various industrial sectors, which are gradually accessing into our daily life. Electrospinning is a versatile and straightforward approach to fabricate nanofibers with controllable compositions and structures for a wide range of applicationsincluding environment, energy, and biology. Whereas, the current electrospun nanofibrous materials generally suffer from some bottlenecks, involving the brittleness of one-dimensional (1D) ceramic nanofibers, the difficulties in thinning the fiber diameters and optimizing the torturous porous structure of two-dimensional (2D) nanofibrous membranes, and the anisotropic lamellar deposition property of three-dimensional (3D) nanofibrous materials, which have significantly limited their practical applications. Aiming at addressing the current challenges of nanofibrous materials, we carried out a systematic study towards the controllable fabrication of 1D soft ceramic nanofibers ($SiO_2$, $TiO_2$, $Al_2O_3$, etc.), 2D ultra-fine nanonets with extremely small diameter (<20 nm), 2D tortuous nanofibrous membranes with tunable surface wettability, as well as 3D isotropically assembled nanofibrous aerogels. The obtained materials can serve as promising candidates in fields includinghigh-temperature heat insulation, air filtration, waterproof and breathable fabrics, oil-water emulsions separation, and sound absorption, etc.Such progresses are anticipated to pave the way for developing next generation fibers and shapebrighter future for our daily lives.

**Dr., Prof. Liu, Zhongfan**
**Peking University**

Dr. Liu received his Ph.D. in Physical Chemistry at University of Tokyo in 1990, and appointed as Full-Professor, at the College of Chemistry and Molecular Engineering at Peking University since 1993.
Dr. Liu is academician of the Chinese Academy of Sciences.
He is a highly decorated scholar with numorous awards, including: Ariyama Kanetaka Memorial Prize for excellent academic achievements, The Japan-China Science and Technology Exchange Association (1992); Cross-Century Talented Scientist Fund, State Education Commission of China (1993); National Excellent Young Fund, Natural Science Foundation of China (NSFC) (1994); Outstanding Young Scientist Award, Hong Kong Qiushi Foundation (1997); Chinese Analysis & Instrumentation Association Science & Technology Award (1st class, 2005); MOE Science and Technology Award (1st class, 2007); National Natural Science Award (2nd class, 2008); Member, Chinese Academy of Sciences (2011); Chinese Chemical Society-AkzoNobel Chemical Science Award (2012); Baogang Outstanding Teacher Award (2012); Outstanding Scientist selected by Ten-Thousand-Talents Program (Six in total), 2013; Fellow, Institute of Physics (UK, 2004); Fellow, Royal Society of Chemistry (UK, 2014); TWAS Fellow (2015); Founder Special Prize (2016); the Lectureship Award of 2016 Colloid and Surface Chemistry Annual Meeting of Chemical Society of Japan (2016); Beijing Municipal Excellent Teacher (2017); ACS Nano Lectureship Award (2018); and many others...

# Graphene Materials: Synthesis Determines the Future

Liu, Zhongfan
Peking University

High quality graphene materials are the footstone of future graphene industry. As experienced in modern carbon fiber industry over last half century, the synthesis will certainly determine the future of graphene materials. Although great efforts have been done on synthesis since the first isolation of graphene in 2004, there still exists a big gap between the theoretical and realistic graphene. For the industry-level applications, one needs to consider the yield and cost issues in addition to purity, layer thickness and uniformity, domain size, lateral size of flakes, and defect density. The graphene synthesis calls for more technological innovations together with fundamental discoveries.

Over last ten years, we have made great efforts on the chemical vapor deposition (CVD) growth of high performance graphene films. We are working along two different directions towards commercial graphene materials. The first direction is the CVD growth of single crystal graphene wafers targeting electronic and optoelectronic purposes. We have realized a pilot level production of 4 inch single crystal graphene wafers using home-made CVD growth system with a capacity of 10, 000 wafers/year. Using CuNi(111) alloy catalyst, we have succeeded in the ultrafast epitaxial growth of 6 inch single crystal graphene wafers with a growth rate of 50 times faster than on Cu(111), indicating the possibility of low cost production of high-quality graphene wafers. The second direction is the CVD growth of large scale graphene film using commercial Cu foil. We developed the first roll to roll continuous CVD growth system for this purpose. The production capacity reaches a level of 20, 000m2/year with a domain size of 10-20μm. The cost has been reduced to 200 RMB/m2. To increase the growth quality, we also developed an A3-size static CVD growth system with a capacity of 10, 000m2/year and an expected domain size of 0.5mm. Superclean graphene is our important contribution to produce high-performance graphene films. During CVD growth, there exists an inevitable contamination of graphene surface arising from amorphous carbon byproduct and not-well developed graphene seeds, which leads to the ordinary "dirty graphene". We have developed three effective techniques to grow the superclean graphene, including Cu-foam-aided growth, post-growth $CO_2$ etching, metal-containing precursors and magic lint roller. Such kinds of superclean graphene exhibited the highest carrier mobility, the lowest contact resistance and sheet resistance, and the highest fracture strength. We designed two types of pilot CVD growth systems based on the $CO_2$ etching technique, A3 size with a capacity of 10, 000m2/year and 300nm x 100mm size with a capacity of 30, 000 pieces/year. Moreover, we have made great efforts on directly growing graphene on traditional glass, optical fibers, glass fibers and sapphire wafers. Such kinds of growth products can be directly used for various applications without involving the difficult peeling off and transfer processes. Super graphene glass, graphene-tailored optical fibers and glass fibers using the direct growth technique have become our important research targets in the last few years. The talk will give a brief overview of our last 10 years studies on the industrial level synthesis and the killer applications of graphene materials.

**Dr., Prof. Poulin, Philippe**
**CNRS – University of Bordeaux**

Dr. Philippe Poulin is CNRS researcher at the Centre de Recherche Paul Pascal in Bordeaux, France. His fields of interest include soft condensed matter, nanostructured and functional materials. He obtained a PhD degree from the University of Bordeaux in 1995. Then he undertook postdoctoral research at the University of Pennsylvania, USA before taking up his current CNRS position. His main achievements include the discovery of novel liquid crystal emulsions, the development of carbon nanotube fibers, and the study of new shape memory nanocomposites. P. Poulin is currently working on the physical chemistry of carbon nanotubes and graphene and on their behavior in polymers and complex fluids. The main targeted applications include energy conversion, dielectric and electrostrictive materials used as actuators, sensors or energy harvesting devices. P. Poulin is author or co-author of more than 170 publications in peer reviewed journals, and co-inventor of 18 patent applications. He has given more than 80 invited talks at national and international conferences. He is vice-director of the Centre de Recherche Paul Pascal and member of the board of reviewing editors of Science.

# Wet-Spun Nanocomposite Fibers

Poulin, Philippe
CNRS – University of Bordeaux

Current industrial synthetic fibers are generally made of neat polymers. These fibers have been optimized over several decades. They display high performances but with little room for further improvements. Enhanced properties and new fiber functionalities call for innovative approaches. In this context, nanocomposite fibers made of polymer and nanoparticles provide an almost unlimited field for the development of new fibers. We will present here recent work on wet-spun fibers that contain inorganic nanoparticles including carbon nanotubes, graphene, ZnO, BaTiO3 or TiO2. Wet-spinning, by contrast to melt-spinning, allows high amounts of nanoparticles to be embedded in the core of the nanocomposite fibers; offering thereby the opportunity for large modifications of properties and new functionalities. We will stress the technical challenges of the approach that involves a fine control of the solution thermodynamics and rheology for efficient fiber wet spinning. We will show that wet-spun nanocomposite fibers can display giant toughness, and also original shape memory behavior with the possibility to make smart textiles with shape changing capabilities associated to temperature memory and large energy density. Nanocomposite fibers also provide new opportunities to develop conductive fibers for sensing, actuating and energy harvesting applications. They can be used in particular for electrodes of biomedical sensors and for enzymatic or microbial biofuel cells. Last, we will discuss new developments in the use of biosourced polymers for nanocomposite fibers. Most synthetic fibers are today made of petroleum based polymers. Making future fibers from biosourced polymers is a critical challenge for a more sustainable fiber industry. In this context, the use of lignin and cellulose is particularly promising. We will show that the addition of carbon nanoparticles in such biobased fibers can efficiently improve their structure. The concept of nanocomposite fibers appears therefore as a promising route to shape a better and more sustainable future.

**Dr., Prof. Kim, Jaehwan***
**Inha University**

Jaehwan Kim received his B.S. degree in Mechanical Engineering from Inha University in 1985, M.S. degree in Mechanical Engineering from KAIST in 1987, and Ph.D. degree in Engineering Science and Mechanics from The Pennsylvania State University in 1995. In March 1996, he joined the Department of Mechanical Engineering at Inha University, Korea, where he is Inha Fellow Professor. He is an associate editor of Smart Materials and Structure as well as Smart Nanosystems in Engineering and Medicine, and editorial board member of International Journal of Precision Manufacturing and Engineering, and Actuators. He has been the director of Creative Research Center for EAPap Actuator funded by National Research Foundation, Korea. Dr. Kim is a fellow of The Korean Academy of Science and Technology, and Institute of Physics. His research interests are smart materials, structures and devices, electroactive polymers, cellulose smart materials, microwave power transmission, biomimetic actuators, biosensors and flexible electronics.

Prof. Kim has first discovered cellulose as a smart material, which can be used for sensors, actuators and electronic materials. His research interests are smart materials, structures and devices, electroactive polymers, biomaterial based smart materials, cellulose, power harvesting, biomimetic actuators, biosensors, tactile sensors and flexible electronics. He has published more than 200 prestigious journal papers, 220 international conference papers, and 30 patents.

# Environmental-Friendly, Strong and Tough Long-Fiber Fabrication by Using Nanocellulose

Kim, Jaehwan
Inha University

Nanocellulose is known to have high specific strength and elastic modulus. Nanocellulose can be mainly divided into cellulose nanofiber (CNF) and cellulose nanocrystal (CNC), which have merits in terms of renewability, high specific strength, high specific elastic modulus, lightweight, abundance, low price and low thermal expansion coefficient. Young's modulus of CNF is in the range of 88-110 GPa and its tensile strength is 1-3 GPa depending on resources. However, its length is too short to form long-fiber. Recently, CNFs were formed to a long-fiber by means of wet extrusion. Its elastic modulus, 22.5 GPa, is lower than that of cellulose nanocrystal and more or less than those of natural fibers like flax, jute, hemp and ramie fibers. The natural fibers are not enough to support the goal of structural composites because they are too short to form long-fiber and have natural variability which limit the mechanical strength. Thus, nanocellulose long-fiber of high elastic modulus should be developed to meet the goal of the proposed research.

This paper presents the fabrication of cellulose long fiber (CLF) which is environmental-friendly, strong and touch. CLF can be made by spinning or aligning nanocellulose. However, CLF for structural composites should have high strength and stiffness as well as toughness because toughness is also an important factor for structural composites to endure under external loads. It is known fact that deceasing the natural fiber size into CNF can improve mechanical properties of CLF. Furthermore, once the CNFs are aligned, then the bond sites between them could be drastically increased. Thus, an efficient way of CNF alignment is essential for fabricating strong and tough CLF. This presentation summarizes current state of the art of CLF fabrication and show recent progress on this area, including CNF isolation, wet spinning, stretching, aligning and drying. Several methods to align the CNFs by applying electric field, magnetic field and mechanical stretching are presented. By securely bonding the well aligned CNFs with drying, a strong and tough structure of CLF can be made.

**Dr., Prof. Voit, Brigitte**
**Leibniz Institute of Polymer Research Dresden**

Brigitte Voit received her PhD in Macromolecular Chemistry in 1990 from University Bayreuth, Germany, in the field of photoactive polymers. After postdoctoral work in 1991/1992 at Eastman Kodak in Rochester, USA, in the field of hyperbranched polymers, she joined Technische Universität München continuing the work on dendritic polymers and she received her habilitation degree in Macromolecular Chemistry in 1996. In 1997 Brigitte Voit was appointed head of the Institute of Macromolecular Chemistry at the Leibniz Institute of Polymer Research (IPF) Dresden, as well as full professor for "Organic Chemistry of Polymers“ at Technische Universität Dresden (TUD). In addition, since 2002, she is heading the IPF Dresden, a large public research institute hosting about 500 people, as Scientific Director.
Her major research interest is in the synthesis of new functional polymer architectures by various synthetic approaches covering topics like dendritic polymers, hydrogels, functional block and graft copolymers, as well as biofunctional polymers and functional polymers for optoelectronic applications. For the period 2014/2015 Brigitte Voit had been president of the European Polymer Federation and in 2015 she was awarded with membership in acatech (Akademie der Technikwissenschaften). She received the Sächsischen Verdienstorden (Saxon Order of Merit) in 2017 and the Hermann Staudinger Award of GDCh in 2018.

# Polymers Designed for Flexible and Opto-Electronics

Voit, Brigitte
Leibniz Institute of Polymer Research Dresden

The organic/flexible electronics industry continues to demand new innovative polymeric materials. The careful macromolecular engineering of conducting and semiconducting polymers and other organic materials with special optoelectronic properties but also features of system integration of these materials are highly important since they determine significantly the final performance of the device. e.g. in organic field effect transistors (OFETs) or in organic light emitting diodes (OLEDs).

We developed hyperbranched polyvinylsulfides through thiol-yne addition of $B_3$+$A_2$ monomers which proved to be excellent candidates for high refractive index (HRI) materials (up RI=1.79), well suited for 1D planar all-polymer photonic crystals and efficient light out-coupling from OLEDs. Furthermore, cyclic oligocarbazoles with benzophenone substituents show excellent *Thermally Activated Delayed Fluorescent* (TADF) properties and are very promising, easy to prepare and highly efficient emitter materials for futures OLEDS .

We will further demonstrate examples of newly developed polymers of high charge mobility as very promising, printable and stable active materials of printed organic field effect transistors (OFETs) and in other flexible electronic applications. Here we emphasize how import it is to also control the chain growth and macromolecular architecture in the area of polymeric semiconductor. We could demonstrate by Kumada Catalyst-Transfer Polycondensation (KCTP) the controlled preparation of dithienosilole (DTS) based conjugated polymers as well as block copolymers with thiophene. The chain-growth polycondensation process could be transferred to n-type naphthalene diimide based conjugated polymers via an unusual anion-radical mechanism which led to ultra-high molar mass conjugated polymers with excellent performance in all-polymer printed OFETs. By a similar process it was also possible to control the number of repeating units and donor and acceptor end groups of thiophene oligomers, and DNA segments could be conjugated to suitably end-group modified controlled polythiophene.

But also step growth procedures have been further optimized. Thus, donor-acceptor polymers have been fine-tuned in their electronic properties and microstructure by molecular composition design. Charge mobility in diketopyrrolopyrrole (DPP) polymers could be significantly increased by improving the synthetic procedures and film processing conditions. With those fully printed all-polymer OFET with charge mobility of 0.42 cm2/Vs have been prepared, and, based on working inverters, the very challenging fabrication of a flexible, fully printed, all-polymer ring oscillator was successful. Furthermore, highly efficient new dopants have been developed for organic DPP polymers, which also allow a spatially defined interfacial doping.

**Dr., Prof. Sun, Gang**
**University of California, Davis**

Dr. Sun studies and develops fibrous materials for functional, protective and eco-friendly clothing. Using advanced polymer chemistry, Sun and his team aim to develop textiles that give wearers, particularly in the medical field, maximum protection against biological and chemical hazards, fire and heat, and allow for the integration of devices such as sensors and smart technology. Sun also studies nanofibers and polymeric materials for applications in biological protection and food safety, as well as for use in pesticides and antibiotics.

Professor Sun has won many awards including: CAREER Award, National Science Foundation (NSF); Director's awards and Circle of Excellence, US National Textile Center (NTC); J. W. Weaver Best Paper Award, AATCC; Chang Jiang Scholar Professor at Donghua University, China Ministry of Education. And he is currently appointed as the Associate editor of Journal of AATCC Research and the Associate editor of Fashion and Textiles.

# Biological and Chemical Sensors Made from Microporous and Nanofibrous Membranes

Sun, Gang
University of California, Davis

Nanofibrous membranes with ultrahigh surface area and highly interactive and selective media to chemicals and biological agents were developed as sensor materials for toxic chemicals. Surface modifications of nanofibers in the membranes were designed and implemented with desired selectivity and reactivity to target chemicals. The surface modification methods employed in the development of the sensors include: 1) physical addition of a layer of organic non-volatile agent with selective adsorptive functions onto surfaces of nanofibers; 2) chemical incorporation of functional groups onto surfaces of nanofibers to provide self-catalytic functions on the membranes; and 3) chemical incorporation of enzymes onto the surfaces of nanofibers to provide specific bindings with biological agents. With a combination of the above mentioned approaches, highly sensitive and selective colorimetric and electrical sensors for vaporous pesticides (fumigants) in air, antibiotics in foods, and pesticides in water were developed based on the microporous and nanofibrous membrane materials. This presentation will discuss theoretical and experimental strategies of designing and developing nanofibrous membrane based sensor materials with improved selectivity and sensitivity.

## Session A
**KEYNOTE SPEECH**

# Programmable shape memory polymer composites and 4D printing: from materials, structures to applications

Jinsong Leng
Harbin Institute of Technology (HIT)

Shape memory polymers (SMPs) and their composites (SMPCs) are capable of changing shapes or sizes when exposed to an external stimulus. Owing to the advantages including fast response, light weights, long lifetime, low cost, high resilience and easy processing, SMPs are the promising materials for smart materials and structures in various areas. With addition of time as the fourth dimension, the 4D printing technology allows the fast development of "living" adaptable structures that enables the growth opportunities in aerospace, medical science and other applications.

Our group is focusing on the advanced 4D printing of SMPs and SMPCs. We have developed a series of 4D printed SMPs and SMPCs structures including tracheal stent, vascular stent, bone stent, and space deployable structures, showing the shape morphing in response to environmental changes. 4D printed structures are capable of remote actuation with magnetic and electricity filed by incorporating of functional fillers in defined regions. Multistage, reversible, programmable structures are also 4D printed and systematically investigated. Moreover, 4D printing technology is developing and will lead to revolutionary in several important fields. Furthermore, many opportunities and challenges in the smart materials and structures are addressed.

# Effect of solution concentration on tensile and creep behavior of gel-spun UHMWPE fibers

Junrong Yu, Hongqiu Wang, Zuming Hu
Donghua University

Solution concentration is a key factor that significantly influences the tensile and creep properties of gel-spun ultra-high molecular weight polyethylene (UHMWPE) fibers. In this study, UHMWPE fibers were prepared by gel-spinning with different concentration of 8~18 wt% UHMWPE solution, followed by extracting and three stage ultra-drawing. The viscosity-average molecular weight (Mη) and maximum drawing ratio of UHMWPE gel fibers were measured. The thermal and crystallizing properties of different concentration UHMWPE fibers were characterized. The morphology structure of different gel fibers were observed. The tensile and creep behavior of resultant UHMWPE fibers were tested. The results demonstrated that the increasing concentration of gel-spinning solution resulted in high loss of UHMWPE Mη and low drawability of resultant gel fibers. The network structure of the gel fibers became denser with increasing of UHMWPE concentration, while the melting temperature, crystal sizes and crystallinity of UHMWPE fibers decreased. Thus the tensile properties of resultant UHMWPE fibers decreased and their creep behavior became poor.

# Perfection of preliminary composite materials performance test standards

Xiangqian Li
Commercial Aircraft Engine Co. Ltd, Aero Engine Corporation of China (AECC CAE)

With the rapid development of high-performance composite materials, the current composite material performance testing standards have been difficult to meet the needs of new design and new process development. This paper proposes an improved and perfect background and perfect direction for a number of current composite material performance testing standards, mainly related to the following standard improvement of the test: (1) the strength dimension effect test standard of tensile strength of composite materials; 2) GIc and GIIc performance test standards between +theta/-theta laminates; (3) GIc fatigue test standards for load control; (4) high toughness GIIc measurement standards after interlaminar toughening; (5) high measurement Improved flat tensile toughness G of ductile composites; compact tensile and compression (CT, CC) test standards; (6) standard for measuring fracture toughness in 33 direction; (7) G12 test standard for plane longitudinal and transverse shear modulus; ) s13 shear strength test standard; (9) short beam girder shear test standard correction for continuous slanting angle ply; (10) virtual test of ply and woven composite properties.

# Ultra-fast, energy-efficient, and mechanically robust nano-welding by localized electrothermal shock towards hybrid nano/macroscale fiber composite manufacturing

Kun Fu
University of Delaware, US

Scaling up of carbon nanomaterials mostly reply on creating strong bonding with macroscopic bulk materials and structures to achieve hybrid properties and enhanced performances, and good durability and high mechanical robustness are required to cross multiple length scale in the hybrid nano-macro materials and structures towards practical applications. However, promoting good contact between carbon nanomaterials and macroscopic bulk structures is a huge challenge. In this work, we demonstrated an electrothermal shock method of utilizing the high temperature joule heating enabled by the high contact resistance of carbon nanostructures and the rapid thermal response of carbon nanomaterials to locally heat and form a strong and robust mechanical bonding between carbon nanomaterials and host materials. We selected CNT/glass fiber system as the model study to study the feasibility of electrothermal shock. The resultant CNT network coated on glass fiber functions excellently as nano-heater to melt glass at contact area with a heating rate over 10000C/s, where the melted glass could physically anchor CNT network, forming a strong bonding between glass fiber and CNT and promoting toughness of glass fibers. The design principle based on the model study of CNT/glass system could be also extended to other a series of carbon nanomaterials and macroscopic host materials with apparent melting temperatures. The excellent performance and potentially lower cost of the electrothermal shock technology offers a continuous, ultra-fast, energy-efficient, mechanically robust, low-cost, and roll-to-roll process as a promising heating solution for hybrid nano/macro-manufacturing across multiple length scales.

# Interface-engineered nanofibrous heterojunctions as high-efficiency electrocatalysts toward ambient ammonia synthesis

Yi-Tao Liu
Donghua University

Ammonia plays an important role in enriching our lives since it vitally participates in modern agricultural, pharmaceutical, and plastic industries. The fixation of nitrogen in the atmosphere is essential to the ammonia synthesis, which is a challenging task since it requires an extremely high energy to break the inert N≡N bond (940.95 kJ mol−1). As such, the century-old Haber–Bosch process for the artificial nitrogen fixation depends on harsh reaction conditions (300–550 °C, 200–350 atm), which lead to huge energy consumption. Besides, the excessive consumption of fossil fuels associated with this process leads to serious CO2 emissions. In this circumstance, the electrochemical nitrogen fixation stands out as one of the most promising candidates, since it can realize nitrogen fixation under ambient conditions.

To enable efficient nitrogen fixation, highly active electrocatalysts are required, which are mostly noble metals such as Au, Pt, and Ru. Recently, the research focus turned to the development of low-cost electrocatalysts composed of nanostructured semiconductor materials, since their unoccupied d orbitals were able to accept the electrons of nitrogen, and thus, break the highly symmetric electronic cloud of the N≡N bonds. However, nanostructured semiconductor materials suffered from low activity and Faradaic efficiency due to two inherent deficiencies, i.e., strong agglomeration tendency and poor carrier mobility. Therefore, several studies employed carbonaceous materials as matrices to support these nanostructures, thus avoiding their agglomeration and improving their conductivity. However, carbonaceous materials were electrocatalytically inactive, so they would weaken the electrocatalytic performances of the resulting hybrid electrocatalysts.

In our work, various ceramic nanofibers were employed as active matrices for the first time to stably confine nanostructured semiconductor materials, thus forming double-active nanofibrous heterojunctions. Through coordination-based interfacial engineering, intimate coupling was created between the two components, which enabled easy carrier transport in between, and thus, fast reaction kinetics at the heterointerface. In contrast to those inactive matrices such as carbonaceous materials, the ceramic nanofibers could contribute substantially to the electrocatalytic performance due to their decent electrocatalytic activities. Profited by this interfacial design, the double-active nanofibrous heterojunctions served as high-efficiency electrocatalysts toward ambient ammonia synthesis, delivering high ammonia yields and Faradaic efficiencies comparable to, or even better than those of noble-metal-based electrocatalysts.

# Macroscopic interfaical supramolecular assembly

Mengjiao Cheng
Beijing University of Chemical Technology

The integration of fibers with other materials could normally lead to reinforced performance of the composites depending on good alignments of fibers and strong interfacial binding of heterogeneous materials. Most techniques rely on surface functionalization of fibers to increase interfacial compatibility and subsequently use traditional spinning methods to align fibers. Insight into enhancement of interfacial adhesion and improvement of ordered degree in fiber alignment are both lacking. Herein, we propose a new concept of macroscopic interfacial supramolecular assembly (MISA), which enhances interfacial non-covalent interactions by more than one magnitude between rigid surfaces through multivalency. Moreover, we have developed a bottom-up assembly method based on MISA by combining magnetic-field-assisted alignment of building blocks with enhanced interfacial assembly to construct ordered composite structure, which is potentially to exhibit high strength and toughness.

# In situ nanofiber-reinforced molecular composites

Qingbao Guan, Zhengwei You
Donghua University

Although the concept of molecular composites(MCs) is very promising, there are major obstacles arisingfrom the immiscibility of the rigid-rod with the random-coilpolymers. Here, we developed a novel method for fabricatingan in situ reinforced MC system with nonequilibrium self-assembled nanofibrous structures based on bisphenol A epoxyresin, 4, 4′-diaminodiphenylsulfone, bismaleimide, and apolyphenylene ether (PPO) oligomer. A variety of spectro-scopic and morphological techniques were used to probe thestructural evolution from the emergence of nanofibrils, togrowth and aggregation of nanofibers, and then to theformation of in situ reinforced MC with strong interfacialinteractions. The in situ nanofibers within the polymer matrix could be formed by the polymerization force extruding the PPO phase through the interspaces within the simultaneous interpenetrating network polymers during the cure process of the thermosetting resin system. Compared to the control sample, the in situ nanofiber-reinforced MC exhibited better thermal properties and flame retardancy. In particular, the obtained MC showed a significant improvement in glass transition temperature and mechanical properties, which were mainly attributed to the restriction of high thermal stability of PPO on the segmental motion of polymer chains, the toughening and reinforcement behaviors of PPO nanofibers on the matrix, and the chemical interaction at the PPO/matrix interface.

1. Guan, Q.; Yuan, L.; Gu, A.; Liang, G., Fabrication of In Situ Nanofiber-Reinforced Molecular Composites by Nonequilibrium Self-Assembly. ACS Appl. Mater. Interfaces 2018, 10 (45), 39293-39306.
2. Guan, Q. B.; Norder, B.; Chug, L. Y.; Besseling, N. A. M.; Picken, S. J.; Dingemans, T. J., All-Aromatic (AB)n-Multiblock Copolymers via Simple One-Step Melt Condensation Chemistry. Macromolecules 2016, 49 (22), 8549-8562.
3. Guan, Q.; Norder, B.; Dingemans, T. J., Flexible all-aromatic polyesterimide films with high glass transition temperatures. J. Appl. Polym. Sci. 2017, 134 (18), 44774.

# Recent research development in fibre reinforced polymer composites for engineering applications

Alan Kin Tak Lau
Swinburne University of Technology, Melbourne, Australia

Since the early of 50's century, carbon fibre reinforced polymer composites have always attracted much attention by the engineering industry owing to their high specific strength-to-weight ratio, ease for producing products with a complex shape by one-stop moulding process, non-corrosive properties and their embedability for placing tiny sensors and actuators to make them multifunction-able and self-defect detectable. However, in the early stage, the price of fibre was a major concern, which restricted the development of this high-strength material only in military and space applications. Until recently, technologies for mass production of these composites have become mature, which greatly lower the prices, to an acceptable limit. Therefore, many industries have started adopting them for both high-end and low-end products. For examples, the use of carbon fibre composites to strengthen civil infrastructures to prolong their service life, or to replace old structures, to make them more light and seismic resistance have been found in United States and Japan. The development of new aircraft fuselage and wings by using hybrid composites can greatly reduce their deadweight, which thus also reduce the fuel use to minimize the production of greenhouse gas from the exhaust of engines. Domestic products including sport utilities, interior components of cars and exterior structures of boats are very common to be seen in everywhere. Although the popularity of using composites has increased in recent years, many inherent problems still exist that may affect the safety level of using them for real-life applications. In this lecture, the recent trend of applications of engineered fibre composites will be reviewed and their problems are also highlighted and discussed with audiences.

# Finite element analysis of large deformation of chemical fiber reinforced braking rubber hose

Liqun Zhang
Beijing University of Chemical Technology

Rubber hose for automobile hydraulic braking system is in lamination structure usually composed of three rubber layers and two chemical fiber braided layers. The chemical fiber braided layer woven with wrap and fill exhibits obviously anisotropic mechanical response, and the rubber layer shows hyperelastic and viscoelastic mechanical response. This makes the high-precision finite element analysis of large deformation for rubber hose difficult. In this work, we present the analytical process of how to model the large deformation of chemical fiber reinforced braking rubber hose with high precision. Firstly, the reliable hyperelastic and viscoelastic constitutive model for rubber materials were established, and the effective mechanical properties of the fabric braided layer treated as an orthotropic material were predicted by representative volume element theory and superposition principle based on JR Cho's work. Secondly, the relations among processing parameters (pitch, number of bundles within a pitch, number of cords within a bundle and helix angle), fiber material parameters and geometric parameters of the fabric braided layer were established. We then developed a Graphical User Interface (GUI) code using Python to accelerate modeling and analysis processes. Lastly, the torsion deformation simulation analysis and experimental comparison were carried out to verify the analysis method was reasonable and reliable. This work can provide fast and efficient theoretical guidance for the structure and material design of high-performance braking rubber hose.

# Thermoplastics reinforced by self-welded short carbon fibers: a nanoparticle-promoted structure evolution

Guozhang Wu
East China University of Science and Technology

Fiber-reinforced polymer composites (FRPCs) have gained wide applications in automobile, aircraft and electronic devices due to their superior strength to weight ratio, high fracture toughness, excellent anti-corrosion and ease of processing. With an annual production of millions of tons, the present state of the art in FRPCs critically limits the intrinsic versatility of fibers. For example, despite of owning high heat resistance of fibers, the heat deflection temperature of FRPCs is inevitably determined by the polymer matrix. Many filaments including carbon nanotubes, ceramic whiskers and metallic fibers behave themselves excellent thermal and electrical conductivity. However, due to the contact resistance between the filaments, present technologies may fabricate FRPCs with conductivities no more than 1% of the fiber itself. In this study, we show a low-cost and widely applicable strategy for "self-welding" the randomly dispersed short fibers to be a three-dimensionally continuous scaffold in the polymer matrix. This has been achieved by adding a small amount of "solder", a third component capable of preferential segregation at the junction point of two fibers. To make the "solder" preferentially segregate on the correct position during compounding and forming process via traditional polymer processing machines, heterogeneity in chemistry or physics of the fiber surface and control of the interfacial affinity and rheological difference between the "solder" and the matrix polymer have been carefully investigated.

# Preparation of thermally conductive silicone rubber composites

Yong Zhang, Jianan Song
Shanghai Jiao Tong University

Thermally conductive silicone rubber (SR) composites have drawn great attention due to their application in electronics to solve heat accumulation caused by the power density increase of electronics. The thermal conductivity of SR composites could be influenced by filler alignment, dispersion and interfacial interaction between filler and matrix. Herein, we introduced facile methods for increasing thermal conductivity and decreasing interfacial thermal resistance. (1) Highly oriented graphene/SR composites were fabricated by spin-assisted layer-by-layer assembly, and exhibited highly ordered lamellar structure and high in-plane thermal conductivity at a low filler loading. (2) Alumina/SR composites with three-dimensional scaffolds were prepared through foaming and subsequent infiltrating. The alumina particles were arranged in the scaffolds and thus formed 3D filler network, leading to the increase of the thermal conductivity. (3) The effect of the interfacial layer on the interfacial thermal resistance and thermal conductivity of triple-layered composites were studied. The different layers in the layered composites were assumed as the filler, matrix and interfacial layer according to their different thermal conductivity. The big difference of thermal conductivity between filler and matrix would lead to a prominent effect of the interfacial layer. (4) Three new grafted siloxane copolymers were synthesized by the reaction of acrylic monomers with SR and used to improve the dispersion of filler and control the interfacial interaction between filler and matrix, and the effects on the mechanical properties and thermal conductivity of SR composites were studied. The structure and properties of highly thermally conductive SR composites were studied.

# Natural silk reinforced epoxy resin composites: design, fabrication and mechanical properties

Juan Guan, Kang Yang
Beihang University

Natural silks are a family of fibres produced by various animals such as spiders and silkworms. These fibres have varied degrees of structural order, and the tailored structure leads to synergistic elastic-plastic deformation mechanisms for balanced strength and toughness as structural materials. We choose natural silks to reinforce a widely used matrix polymer, epoxy resin, and create a distinct series of composites for optimal toughness and impact-resistance.

In the series, two commercially available natural silks from common Bombyx mori (Bm) and semi-wild Anthearae pernyi (Ap) silkworms are selected. The composites are named as silk fibre reinforced plastics or SFRP. The following findings are summarized based on our earlier work: (1) the fibre volume fraction has a clear effect on the mechanical performance of Bm-SFRP, and as high as 70 vol.% fibre volume fraction can be realized for Bm-SFRP; (2) with >50 vol.% silk, the impact resistance of SFRPs can be dramatically lifted; (3) >50 vol.% Ap-SFRP displays superior breaking energy absorption, impact strength and sub ambient temperature resilience; (4) flax fibre as a high-performance plant fibre is applied to modulate the mechanical performance of SFRP; (5) high-stiffness and strength carbon fibre is hybridized with silk to create new sets of mechanical properties in composites. We also study the molecular interactions of silk protein polymer and epoxy resin to deepen the understanding on the silk-epoxy resin interface in the composite. In the future, we expect to use these SFRPs for secondary structural / structural applications and impact-critical scenarios.

# Large-scale preparation and application of high performance polyimide fiber

Jie Dong, Xin Zhao, Qinghua Zhang
Donghua University

High strength and high modulus polyimide fibers are a type of novel high-performance organic fiber with an initial modulus higher than 100 GPa and extremely high tensile strength over 4.0 GPa realizing broad applications in the fields of advanced composites, high-temperature dust removal, special protection, insulating materials, aerospace and atomic energy industries. Herein, an integrated fabrication process involving the "reaction-spinning" followed by the thermal imidization is reported as a typical synthetic route for the industrialization of high-performance polyimide fibers. In special, the relationship between the structure (molecular structure, morphology, interchain association, aggregation structure) and the performance (mechanical property, thermal stability, dielectric behavior as well as UV resistance) of polyimide fibers is investigated extensively. In addition, the applications of several commercialized polyimide fibers are presented for a better understanding this new high-performance polymer materials.Polyimide fiber and color yarns prepared by the "reaction spinning" method and the comparison of specific strength and modulus of PI fibers with other polymeric fibers

# Multi-scale modification of carbon fiber by in-situ growth of clickable polymer micro-/nano-particles

Lianying Liu, Huimin Liu, Xingyu Jiang, Shuai Wang, Wantai Yang
Beijing University of Chemical Technology

Carbon fiber reinforced polymer composites (CFRPC) are widely used in aerospace, high-speed rail, automobile and other fields, due to their advantages of light weight, high strength, good chemical stability, multiple designability, etc. The interphase between carbon fiber (CF) and matrix plays an essential role in determining the performances and applications of CFRPC. However, CF surface shows chemical inertness and poor interfacial compatibility with matrix. This finally affects the mechanical properties and service life of composites. Therefore, various methods have been developed to modify the surface of CF, such as the introduction of small molecules, branched or hyperbranched macromolecules, GO nanosheets, carbon nanotubes, $SiO_2$ or $TiO_2$ nanoparticles and other multiscale micro/nano structure on CF surface. But, few attentions have been paid to grow polymer micro-/nano-particles on CF, probably due to the lack of efficient techniques used under mild conditions.

It is known that click reactions can be carried out fast, under mild conditions, and using wide variety of available monomers. In recent years, heterogeneous click-polymerizations including thiol-isocyanate or epoxy click-disperse polymerization, etc., have been used to prepare functional polymer microspheres.

In our work, to achieve multi-scale modification of CF, and thus to improve the interfacial strength between CF and polymer resins, we propose a simple and fast method to in-situ grow polymer micro-/nano-particles by click dispersion polymerization of thiol-isocyanate or epoxy on CF. Surface roughness of CF is improved and functional groups, such as -SH, -NCO, $-NH_2$ or fluorescent groups, are introduced on CF at the same time. As a result, chemical bonds and mechanical interlock interaction between CF and matrix are formed. This effectively improvs the interfacial strength. Furthermore, with the aids of in-situ grown fluorescent particles, the interface crack of composite materials can be marked and detected. Our work provides a new method for multi-scale modification of CF surface, controlling interface performance.

# Tuning the interfacial property of fiber reinforcement composite by constructing 3D structure on fiber surface

Dan Pan, Siqi Liu, Long Chen, Baozhong Sun
Donghua University

Many technical problems, such as weak bonding between fiber and base or poor fiber dispersion in base, constrain the application of fiber reinforcement composite (FCR) materials. The surface structure of fiber plays a key role in tailoring the surface properties of fiber and interfacial properties of FCR materials. In this research, the surface structure was constructed based on tuning the micro-rheological behavior of second phase in polymer blend during melt spinning. Some post procedures were introduced to help change the morphology of surface patterns. The surface properties were characterized by scanning electron microscopy and atom force microscopy. The interfacial properties of FCR are studied based on single fiber pull out test and limited element simulation. The relation between surface structure and interfacial properties are also built.

# Surface modification of high performance fibers and its effect on interfacial adhesion of elastomer composites

Ming Tian
Beijing University of Chemical Technology

With the requirements of light weight, high strength, high temperature resistance and long life of rubber products, the demand for high-strength nylon/polyester fiber, aramid fiber and other high-performance fibers are increasing in rubber industry. To give full play to the reinforcing effect of high-performance fibers on elastomer, it is a key to modify the surface of high-performance fibers to improve the interfacial adhesion between fibers and elastomer matrix. This paper focuses on the principle of surface coating and secondary functionalization modification of dopamine or dopamine-like system such as phenol-amine, tannic acid system on high-performance fibers and their effects on interfacial adhesion of elastomer composites. The research progress of replacing the traditional toxic resorcinol-formaldehyde-latex (RFL) impregnation system by using new environmentally friendly ones is introduced. As an example, the research progress of the application of aramid fiber in some special and large rubber conveyor belts is introduced.

# Preparation and application of high performance carbon fiber

Anqi Ju
Donghua University

High performance carbon fiber, an ideal reinforcing material for advanced composite materials, has been extensively applied in high-tech aerospace, defense areas, and civil engineering. New bifunctional comonomer were synthesized and used as comonomer to prepare acrylonitrile copolymers. The polymerization kinetics and stabilization of acrylonitrile copolymers were studied in detail. The results show that the stabilization of acrylonitrile copolymers was significantly improved by the synthesized comonomer with lower initiation temperature, broadened exothermic peak and activation energy. Ultra-high molecular weight of acrylonitrile copolymers was prepared by aqueous suspension polymerization, which is beneficial to prepare high performance carbon fiber.

# Nanofibrous scaffolds for tissue engineering

Xiaoran Li
Innovation Center for Textile Science and Technology, Donghua University

Nanofibrous scaffolds with a variety of structure and architecture have been widely employed in tissue engineering owing to better mimicry of structure and organization of the native extracellular matrix. In addition, the functionalized nanofibrous scaffolds would provide favorable microenvironment for tissue regeneration. Stem cell therapy shows great potential in regenerative medicine. Regulation of stem cell fate remains a huge challenge. In our study, we developed functional nanofibrous scaffolds to provide space and cues for directing stem cell migration and differentiation. We have fabricated radially aligned electrospun nanofibers immobilized with continuous gradients of stromal-cell-derived factor-1α (SDF1α). The SDF1α gradient scaffolds directed and enhanced migration of neural stem cells from the periphery to the center along the aligned electrospun fibers, showing great potential in guiding endogenous neural stem cells to the lesion site during repair of spinal cord injury. As another example, we have developed 3D ceramic fibrous scaffold constructed by intrinsically rigid, structurally flexible electrospun $SiO_2$ nanofibers with chitosan as bonding sites. The 3D scaffolds induced human mesenchymal stem cell differentiation into osteoblasts. The superelastic scaffolds were able to self-fit to mandibular defects in rabbits via minimally invasive surgery, and promote bone formation in rat calvarial defect.

# 4D printing of programmable shape memory structures for biomedical applications

Fenghua Zhang, Yuliang Xia, Yanju Liu, Jinsong Leng
Harbin Institute of Technology

Four dimensional (4D) printing technology is a novel method to manufacture the complex structures combining smart materials. 4D printed structures have a wide range of potential applications in aerospace, flexible electronic devices, photoresponsive device, and biomedical science. The advantages of 4D printing is that the printed structures show shape changing behaviors depending on time. Shape memory polymers as a kind of smart material have attracted great attention because of the large deformation, low cost and easy processing. In this work, we presented various 4D printed structures fabricated by biocompatible and biodegradable PLA/Fe3O4 composite filaments. The 4D printed composite structures can be programmed and triggered by magnetic field, showing fast recovery behaviors. Moreover, the surface temperature of 4D printed structures was uniform and low during the shape recovery process. The compress structure can expand to the required shape and size upon remote actuation by a magnetic field. Therefore, the 4D printed PLA/Fe3O4 composite structures have great potential in biomedical applications.

# High performance, multifunctional three dimensional composite structures

Fujun Xu, Kun Zhang, Yiping Qiu
Donghua University

Smart or intelligent textiles are considered as the next-generation wearable electronics. However, in this smart textile systems, one of the key components is the antenna that transmits signals between the functional devices. Therefore, fabric-based antenna plays an important role in wireless body area network systems (WBANs) or other flexible and conformal antenna systems. In this study, we proposed an ultra-low density three-dimensional woven spacer antenna (3DWSA) for the first time by integrating microstrip antenna into 3D woven spacer composites. The 3DWSA substrate had a spacer structure with the air volume fraction of ~75%, rendering a low density (~0.5 g/cm3), dielectric constant (1.75) and dielectric loss (0.0085). Therefore, it showed excellent electromagnetic performance with the gain values of 7.1 dB (single-element 3DWSA) and 9.3 dB (double-element 3DWSA) respectively, which also agreed well with the simulated results. Furthermore, the 3DWSA maintained proper resonant frequency and impedance matching after the impact of 18J, exhibiting excellent structural integrity.

# Preparation of high performance heterocyclic aramid fibers with enhancing compressive strength

Longbo Luo, Xiangyang Liu, Zheng Cheng, Yu Dai
Sichuan University

The heterocyclic aramid has excellent mechanical and thermal properties. However, weak interaction between polymer chains and microfibrils leads to low compressive strength, which limits the application in advanced composite materials. In this paper, we reports three methods to enhance compression strength of aramid fiber. The first one utilizes the reactivity of N-H groups on the benzimidazole ring with crosslinkers to carry out interchain crosslinking; the second one is preparation of the "fence" structure of the fluorinated carbon nanotube and the aramid molecule. The enhanced molecular chain interaction enhances the compressive strength; the third is inspired by the high compressive strength of the three-dimensional support structure in the building. The hyperbranched polyamide modified nano silica is constructed in the aramid fiber body. The compressive strength of the heterocyclic aramid prepared by the above method was increased by 100%, 107% and 114%, respectively, compared to neat heterocyclic aramid. Moreover, the tensile strength of modified heterocyclic aramid is also improved due to crosslinking and nanocomposites.

# Structure and performance changes of polyester fiber under various water drawing conditions

Yang Zhang, Jiaxiong Zou, Yumei Zhang
Donghua University

High-strength polyester fibers are usually prepared by multi-stage, multi-hot drawing methods(1). The hot air (2) was used as drawing medium in the current drawing process, and the premature formation of crystallization during hot air drawing will hinder the orientation of polyester fibers, which further inhibit the formation of crystallization to some extent (3). Existing experiments (4) have investigated the conjugation effect of drawing and water absorption of PET films in hot water baths with the results in changes of Tg and Tc little change of crystallinity.

In this work, water bath was selected as a draw medium to investigate the effect of water bath drawing conditions such as temperature and drawing ratio (λ) on the structure and performance of polyester fibers.

In summary, the PET fiber with low crystallinity and high amorphous orientation could be obtained by drawing in water, which indicates that the molecular chain would be easy to be arranged into a more ordered structure as the PET fiber is subjected to further heat drawing and heat setting. And the PET fiber with less defect structure could be prepared as expected.

# Effect of compatibility on morphology and properties of polypropylene/ coir fiber composites

Chaoqun Li
Hainan University

Polypropylene/ coir fiber composites were prepared through two-step blending by firstly adding compatibilizer and part of polypropylene, then the rest of polypropylene were added. The dispersion of coir fiber, interfacial adhesion between coir fiber and polypropylene were studied using scanning electron microscope, which illustrated the improvement in interfacial interaction after the introduction of compatibilizer, which result in an increase of more than 30% in tensile strength through the compatibilization effect. Besides, the stiffness, thermal stability increased remarkablely in the compatibilized systems compared with neat polypropylene. THe processability of the samples were also discussed. The compatibilized systems showed good comprehensive properties.

# Effect of TiO2 on the spinability, structure and property of polyester industrial yarn

Yan Liu[1], Hong Ji[1], Yumei Zhang[1], Minggen Song[2], Quan Jiang[2]
1.DongHua University
2.Zhejiang Unifull Industrial Fiber Co., Ltd

Nano TiO2 has a good uv reflection function, and has excellent chemical stability, thermal stability, non-migration and strong decoloration and hiding power. With the use of TiO2 high refractive index and uv reflection function, uv-resistant polyester fiber has been successfully prepared. Extinction, resistance to ultraviolet, however, the high strength polyester industrial yarn has not been industrialized, therefore, to study nano TiO2 effect on the properties of polyester industrial yarn preparation and construction, we through (DSC), differential scanning calorimeter rotational rheometer nanometer TiO2 is studied on the crystallization properties and rheological properties of high molecular weight polyester and the effect of the masterbatch to add the preparation of the ultraviolet resistant, high strength polyester industrial yarn. The fiber morphology, mechanical properties and uv resistance were tested.

# Preparation and properties of PVA fibers reinforced with GO nanoribbons derived from unzipping of SWNTs

Xike Xiong, Jun Sun, Lixing Dai
Soochow University

In order to significantly increase the mechanical properties of regular synthetic fibers, effectively, nanofiller-reinforced polymer composite attracted extensive attention in high performance fiber area. In this paper, complete morphology graphene oxide nanoribbons (GO-NRs) were prepared through unzipping single-walled carbon nanotubes (SWNTs) through long-time freeze-drying. GO-NRs obtained from SWNTs (SGO-NRs) could be directly added to poly (vinyl alcohol) (PVA) without any further modification and so it could form uniform dispersions. After that, final fibers were prepared via wet-spinning and hot-drawing.[1-2] SGO-NRs provided abundant hydroxyl so as to form enough hydrogen bonding interaction with PVA chains. As a result, SGO-NRs could not only obviously improve the dispersibility in PVA, but also enhance the mechanical properties of the composites. The tensile strength and modulus of PVA/SGO-NRs composite fiber containing 0.4 wt% SGO-NRs reached 1032 MPa and 10900 MPa, increased by 121% and 24% compared with PVA/SWNTs fiber, and by 200% and 153% with PVA fiber, respectively.

# Highly boosting the interlaminar shear strength of CF/PEEK composites applied in civil aviation

Shu Zhu[1], Dengteng Ge[1], Elwathig A. M. Hassan[1,3], Jianfeng Zhou[1], Huajun Cao[2], Muhuo Yu[1]
1.Donghua University
2.Chongqing University
3.University of Gezira, Sudan

Carbon fiber reinforced polyetheretherketone (CF/PEEK) thermoplastic composite is one of the most promising novel materials applied in civil aviation. However, due to poor wettability and low reactivity, the interfacial adhesions between CF and PEEK matrix are relatively weak, significantly affecting mechanical performances of composites, especially the interlaminar shear strength. This study focuses on improvement of fiber-matrix interfacial interactions for CF/PEEK composites via introducing interfacial layers of polyertherketoneketone (PEKK) on activated CF. This method takes advantage of hydrogen bond between PEKK and activated CF, as well as good compatibility between PEKK and PEEK. As a result, interlaminar shear strength, flexural strength and modulus of CF/PEEK composites increased by 70%, 37% and 48%, respectively. SEM observations on fractured surfaces of composites indicated that dominant failure mechanisms shifted from fiber-matrix debonding to deformation of interfacial layers and breakage in resin. This modification method is easy to handle and can be applied to produce large size of composite materials containing continuous CF or CF fabrics, showing potentials in industrial applications.

# Ultrafast self-healing and injectable conductive hydrogel for strain and pressure sensors

Fengming Ye[1], Meng Li[1], Dingning Ke[2], Liping Wang[1], Yi Lu[1]
1.The Brain Cognition and Brain Disease Institute, Shenzhen Institutes of Advanced Technology, Chinese Academy of Sciences, Shenzhen-Hong Kong Institute of Brain Science-Shenzhen Fundamental Research Institutions
2.Experiment and Innovation Center, Harbin Institute of Technology Shenzhen Graduate School

High-precision sensors are urgently needed to investigate subtle physiological responses against particular stimuli. Injectable conductive self-healing (ICSH) hydrogels are promising candidates to achieve this goal owing to their superior mechanical and electrical characteristics. Here, an interpenetrating polymer network (IPN) of multi-walled carbon nanotube-poly(3, 4-ethylenedioxythiophene)/poly(styrenesulfonate)-polyacrylamide-poly(vinyl alcohol)/borax composite hydrogel is cross-linked in a simplified process. The hydrogel offers an ultra-low Young's modulus, conductivity, high stretchability, rapid self-healing and injectability. A miniature strain sensor fabricated via material injection shows linear stretching responses and can monitor respiration situations of a mouse during different activities. Pressure micro-sensor fabricated through printing–injecting stably monitors sphygmus signals with characteristic peaks from human arteries. Such miniature sensors can contribute to healthcare monitoring and the understanding of the biological basis of a particular behavior or neuropsychiatric disorder.

# New comonomer for polyacrylonitrile-based carbon fiber: density functional theory study and experimental analysis

Huichao Liu, Caizhen Zhu, Jian Xu
Institute of Low-dimensional Materials Genome Initiative, College of Chemistry and Environmental Engineering, Shenzhen University

Carbon fiber is one of the world's three major high-performance fibers, which is known as the "crown jewel" of new materials. Exploiting new comonomers is still required for high performance polyacrylonitrile (PAN) based carbon fiber. In this paper, we have proposed an efficient methodology, combining of theoretical calculation and experimental verification, to develop new comonomer for polyacrylonitrile (PAN)-based carbon fiber. The cyclization energy barriers of PAN copolymers, including comonomers of α-nitryl acrylic acid (IA-NO2), α-amino acrylic acid (IA-NH2), acrylamide (AAM), itaconic acid (IA), and ethylenesulfonic acid (ESA), have been calculated based on the autocatalytic cyclization mechanism using density functional theory (DFT) at B3LYP/6-31 + G (d, p) level. The theoretical calculation indicated that ionic cyclization of nitrile group was more easily initiated by ESA than other comonomers. Correspondingly, the PAN copolymers including comonomers of ESA and IA have been prepared and studied on their properties. The experimental results further demonstrated the P(AN-co-ESA) copolymer had better thermal properties such as smaller cyclization energy, slower heat release rate, and higher char yielding. Therefore, this theoretical calculation combined with experimental verification methodology is a powerful tool for exploiting new comonomer for PAN-based carbon fiber.

# Preparation, stabilization and carbonization of a novel polyacrylonitrile-based carbon fiber precursor

Caizhen Zhu, Huichao Liu, Jian Xu
Institute of Low-dimensional Materials Genome Initiative, College of Chemistry and Environmental Engineering, Shenzhen University

The quality of polyacrylonitrile (PAN) precursor has great influence on the properties of the resultant carbon fibers. In this paper, a novel comonomer containing the sulfonic group, 2-acrtlamido-2-methylpropane acid (AMPS), was introduced to prepare P(AN-co-AMPS) copolymers using itaconic acid (IA) as control. The nanofibers of PAN, P(AN-co-IA), and P(AN-co-AMPS) have been prepared by electrospinning method. The effect of AMPS comonomer on the carbon nanofibers has been studied by scanning electron microscopy (SEM), differential scanning calorimetry (DSC), Fourier transform infrared spectroscopy (FTIR), X-ray diffraction (XRD), thermogravimetric analysis (TGA) and Raman spectrum. The structural evolutions of PAN-based nanofibers were quantitatively tracked by FTIR and XRD during the thermal oxidative stabilization (TOS) process. The results exhibited that P(AN-co-AMPS) nanofibers had the lower heat release rate ($\Delta H/\Delta T$=26.9 J g-1 oC-1), the less activation energy of cyclization (Ea1=26.6 kcal/mol and Ea2=27.5 kcal/mol), and the higher extent of stabilization (Es and SI) during TOS process, which demonstrated that the AMPS comonomer improved the efficiency of TOS process. The P(AN-co-AMPS) nanofibers had the better thermal stable structures. Moreover, the carbon nanofibers derived from P(AN-co-AMPS) precursor nanofibers had the better graphite-like structures (XG=46.889). Therefore, the AMPS is a promising candidate comonomer to produce high performance carbon fibers.

# Self-healing polymeric materials: design, preparation and characterization

Mingqiu Zhang
Sun Yat-Sen University

Self-healing polymers and polymer composites represents a class of materials with built-in capability of rehabilitating damages. Damages that are inevitably generated in materials during fabrication and service can be automatically repaired on a microscopic scale and would no longer develop into macroscopic failure as in the case of conventional materials. The topic has attracted more and more attention in the past few years. The on-going research activities clearly indicate that self-healing polymeric materials turn out to be a typical multi-disciplinary area. The present report briefly reviews the works carried out in the authors' laboratory towards strength recovery for structural applications.
The self-healing methodologies developed to date can be classified as intrinsic and extrinsic according to the method used for delivering the healing components to the target site in the material. Unlike intrinsic self-healing that operates as a result of inter- or intra-macromolecular interactions in the absence of additional healing agents, extrinsic self-healing involves the embedment of a healing agent. Extrinsic self-healing can be more easily realized in commercially available polymers because no structural modification of the matrix molecules is required.

# Unusual NMR methods for advanced fibers and polymers

István Bányai*[1], Mónika Kéri[2], Dávid Nyul[3], Levente Novák[2]
University of Debrecen

NMR spectroscopy plays a dominant role within spectroscopies in many fields of chemistry and related applied sciences. The reason is that the spectra, in general, can be interpreted very easily. The peaks usually are narrow (not overlapped) and the integrated intensity is linearly proportional to the number of atoms. 1H- and 13C-NMR measuring techniques are well elaborated because of their vital importance in organic synthetic chemistry and biochemistry. These advantages override some serious disadvantages such as the low sensitivity of the techniques or the high price of purchase and maintenance of the NMR instruments. The sensitivity, thanks to the new developments, is increasing but unfortunately the price follows it. However, there are NMR techniques hidden from the eyes of applicants although they can give information which cannot be obtained in another way. In this presentation, after a short summary of usual multinuclear and multidimensional NMR methods for fibers and polymers, we will deal in details with NMR cryoporometry, relaxometry and diffusiometry. These methods made possible the comprehensive characterization of porous materials concerning pore size and shape, pore size distribution, pore volume, hydrophilic/hydrophobic properties mostly under the circumstances of application.

NMR relaxometry

The transverse (T2) relaxation time constant of water or organic solvents in porous materials is usually determined at different liquid contents by low field 1H-NMR methodology. This method is based on the difference of relaxation rates of protons of the solvent close to the surface and the bulk phase. In porous media under appropriate conditions the relaxation rate is proportional to the surface to volume ratio. Furthermore, the effect of solvent content on the relaxation rate may be characteristic of the wetting and the hydration mechanism. These experiments can be used to characterize the porosity and pore size distribution of the surface of the fibers.

NMR diffusiometry

The reversible diffusion, diffusion in the absence of chemical potential difference, can be measured by NMR. For determination of the size of nanoparticles “in action” these experiments cannot be circumvented. However the restricted diffusion of the solvent or medium in porous system holds information about pore structure and the average pore size. Since the solvent, containing NMR active nuclei, is present in large concentration, the diffusion coefficient can be conveniently determined with low-field, therefore low-cost NMR. I intend to show applications for porous fibers and polymers underlining the easy industrialization of this method.

NMR cryoporometry

NMR cryoporometry is based on the modified Gibbs-Thomson equation which describes that there is melting/freezing point depression of pure solvent embedded in small pores depending on the pore size. Since a well-designed NMR experiment detects quantitatively the amount of non-frozen solvent we can follow the temperature dependence of the melted to frozen ratio of the solvent, determined by the pore-size distribution. With this method the porosity and the pore size distribution can be determined even for very soft materials, like soft gels and fibers.

The goal of this presentation is to show these NMR methods and their combinations on the application-level and bring closer to each other the NMR spectroscopist and the industrial researcher- in order to characterize the synthesized products.

# Dynamical characterizations of organic-inorganic interface in nano-composite materials using advanced electron microscopy

Hiroshi Jinnai
Tohoku University

The interfacial region between inorganic nano-fillers and polymer matrix in hybrid materials (or alternatively nano-composite materials) is one of the key morphological elements to govern their mechanical properties. The morphological characterization of such interfacial region around fillers, i.e., "interphase region" for short, and its dynamical behavior under deformation has been a subject for many decades.
We have developed a three-dimensional (3D) tomography holder for in-situ tensile deformation for polymeric materials in transmission electron microscope (TEM) [1]. With this new holder, because both ends of (microtomed) thin sections are stretched simultaneously at the same stretching rate, the drifting of field of view for TEM observations becomes minimum. The feature is particularly important especially when the specimen is highly strained. The largest strain achievable with this tensile deformation holder is, in principle, about 50 (when the initial specimen length is 20 μm). Moreover, the tensile holder allows us to tile the specimen up to 75°for tomographic reconstruction. Thus, the present holder is capable of taking 3D images of polymeric specimens under tensile deformation.
The tensile deformation tomographic holder was used to observe deformation processes of rubber composite, in which nano-scale fillers dispersed in polymeric matrix. The morphological changes in rubber composites under stretching, especially the nano-voids formation, as well as the fracture processes were directly observed at nanometer scale, which will be discussed at the conference time.

# Important subjects on prediction of fabric properties from yarn and fabric structure

Masayuki Takatera, KyoungOk Kim, Liu Yang
Shinshu University

Predicting fabric properties using mechanical properties of yarn and fabric structures has been a research theme since 1900s [1]. Peirce proposed a geometry model of fabric structure in 1937[2]. With the development of computer simulation technologies, nowadays fabric geometrical structures can be simulated from yarns. The next step of fabric simulation is using yarn properties and fabric structure to predict the mechanical properties of fabrics. Finite element method and particle method play important roles in the fabric simulation. However, because there are still unsolved important subjects on this theme, the quantitative prediction on fabric mechanical properties using these methods according to yarn properties and fabric structures is limited. In this study, we show the unsolved issues surrounding this topic, and discuss the necessary matters of fabric characteristics.

# Ultra-high-molecular-weight polymers produced by the "immortal" phosphine-based "frustrated Lewis pair" catalyst system

Yuetao Zhang
State Key Lab. of Supramolecular Structure and Materials
Jilin University

The "frustrated Lewis pairs" (FLPs) chemistry has attracted an explosive level of recent interest since the seminal works of Stephan and Erker. In this presentation, a strong organophosphorus superbase, N-(diphenylphosphanyl)-1, 3-diisopropyl-4, 5-dimethyl-1, 3-dihydro- 2H-imidazol-2-imine (IAP3) was combined with a sterically encumbered but modestly acidic Lewis acid (LA), (4-Me-2, 6-tBu2-C6H2O)AliBu2 ((BHT)AliBu2), to synergistically promote the frustrated Lewis pair catalyzed living polymerization of methyl methacrylate (MMA), achieving ultrahigh molecular weight (UHMW) poly(methyl methacrylate) (PMMA) with Mn up to 1927 kg/mol and narrow molecular weight distribution (MWD) at room temperature (RT). This FLP catalyst system exhibits exceptionally long lifetime polymerization performance even in the absence of free MMA, which could reinitiate the desired living polymerization after the resulting system was held at RT for 24 h. These findings not only provide significantly important foundation to the synthesis of UHMW polymer in the future, but also enable us to synthesize advanced polymers such as thermoplastic elastomer or self-assembling polymers by this highly effective and "immortal" catalyst system.

# Biomimetic materials with multiple protective functionalities

Zenghe Liu, Luzhi Zhang, Zhengwei You
Donghua University

Natural tissues possess superior material properties such as self-healing, mechanical robustness, and mechanical gradients that allow organisms to adapt and survive in dangerous environments. Although highly desired, imparting synthetic materials with these biomimetic protective features remains a challenge. Here, the versatile dimethylglyoxime–urethane (DOU) moiety is used to create a multifunctional polyurethane (DOU-PU). The reactivities of DOU including reversible dissociation, metal coordination, photolysis enabled self-healing, high strength and toughness, mechanical gradient formation, and spatially controlled functionalization. By incorporating DOU, a multifunctional protective film is produced with superior resistance to mechanical damage, rapid room temperature self-healing, and anti-counterfeiting features. This super biomimetic film is expected to be very useful for the protection of various types of valuable objects such as electronics, diplomas, currency, and automobiles.

References

1. L. Zhang et al., A Highly Efficient Self-Healing Elastomer with Unprecedented Mechanical Properties. Adv. Mater. 31, 1901402 (2019).
2. Z. Liu et al., Biomimetic Materials with Multiple Protective Functionalities. Adv. Funct. Mater. 29, 1901058 (2019).
3. S. Wang et al., Strong, detachable, and self-healing dynamic crosslinked hot melt polyurethane adhesive. Materials Chemistry Frontiers 3, 1833-1839 (2019).

# Moisture insensitive and superfast ring opening polymerization of N-carboxyanhydrides

Yueming Wu, Runhui Liu
East China University of Science and Technology

Polypeptides have broad applications such as tissue engineering, drug delivery and antimicrobials. It has been more than 110 years since Hermann Leuchs discovered N-Carboxyanhydrides (NCA) for the first time in 1906 and since then NCA polymerization has developed into a classical method for the preparation of polypeptides. However, NCA polymerization is sensitive to moisture and the most widely used primary amine initiator only gives slow polymerization resulting polypeptides in short chain length. Therefore, we developed the LiHMDS-initiated open vessel NCA polymerization that is not sensitive to moisture and can initiate NCA polymerization outside of the glovebox without any protection. The simple operation facilitates large scale and parallel synthesis of polypeptide. In addition, the polymerization can complete as fast as mins and it is suitable to prepare long chain polypeptides that are important for many biological activities and functions.

KEYNOTE SPEECH

# Pleated and rippled sheets of polyamides. The "high temperature" alpha phases of nylons. A structural scheme for the Brill transition.

Bernard Lotz
Institut Charles Sadron (CNRS and Université de Strasbourg)

The structure of nylons rests almost exclusively on sheet structures in which the chains are fully extended: the amide group is coplanar with the C-C-C plane of the aliphatic parts.1 The only known exception is the so-called gamma phase of e.g. nylon 6 in which a twisted conformation exists at the aliphatic part – amide junction.2

In a work under progress, 3 we have considered the possible existence of sheet structures in which such twisted conformations also exist for other types of nylons, and in particular for the series of even-even nylons. Modeling of possible structures has been mainly performed on PA66, but the results are valid for other nylons. Indeed, the "structural signature" (in the form of characteristic hk0 reflections spacings observed in e.g. powder diffraction patterns) does not depend on the length of the aliphatic parts. By analogy with the pleated and rippled sheets of polypeptides introduced in the early 1950s by Pauling and Corey, 4 it is possible to build "pleated" and "rippled" sheets of nylons. The pleated sheets are made of one chain conformation only. In the rippled version, two stems with mirror symmetry conformations alternate in the sheet.

The models thus developed explain a number of earlier observations of nylon crystal structures and transitions. In particular, the pleated sheet is a valid model for the whole family of nylons 2, 2m.5 Also, these pleated and rippled sheets are suitable models for several so-called "high temperature" alpha phases, usually produced after annealing/self-seeding in the melting range and crystallization at slightly lower temperature. For example, the calculated hk0 spacings for the rippled sheet are 4.40Å and 4.25Å. For "a kind of a phase... much different from the room temperature a phase" of nylon 414 the reported hk0 spacings are 4.36Å and 4.23Å.6 Similarly, these new crystal structures provide potential models for a number of unusual crystal morphologies formed after similar thermal treatments (e.g. spherulitic aggregates).7 Twisted conformations apply for all other types of nylons, which opens the way for a more thorough analysis of their complex crystal polymorphism.

Most importantly, the two enantiomeric chain conformations are dynamically interconvertible. This dynamic interconversion (at a ≈10-10 time scale) results in a limited oscillation of the amide groups that maintains the hydrogen bonding within the sheet but generates a significant swivel motion of the aliphatic parts. These are the exact characteristics of the Brill transition.8, 9 The pleated/rippled sheets appear to be a necessary transitional structure for the Brill transition to take place: when both aliphatic segments are "too long" (over ≈12 C atoms), their thermal agitation prevents formation of the pleated sheet – and thus occurrence of the Brill transition - prior to melting.5

# Light-triggered degradation in aliphatic poly(carbonate)s via intramolecular cyclization

D. Kuckling[1], J. Sun[1], W. Birnbaum[1], M.-T. Picker[1], J. Anderski[2], D. Mulac[2], K. Langer[2]
1.Department of Chemistry, Paderborn University
2.Institute of Pharmaceutical Technology and Biopharmacy, University of Muenster

Nanoparticles (NP) offer several advantages as drug vehicles, like a stable entrapment of APIs with a low water solubility or the option of an active drug targeting through surface modifications. Thereby, NP composed of a poly(DL-lactide-co-glycolide) (PLGA) matrix are a prominent representative. However, the in vivo degradation of the PLGA nanocarriers is often difficult to predict.

Aliphatic poly(carbonate)s (PCs) with rapid and controlled degradation upon specific stimulation have great advantages for a variety of biomedical and pharmaceutical applications. Toward this end, we have synthesized new carbonate based copolymers containing multiple photo cleavable groups along the backbone by ring opening polymerization (ROP) and polycondensation.

The light-responsive carbonate monomers (LrM) have been prepared and copolymerized with different carbonate structures yielding UV-light responsive copolymers (LrP). The light-triggered decomposition of LrM and LrP was studied by 1H NMR, IR and UV/VIS spectroscopy, size exclusion chromatography (SEC), as well as ESI-ToF mass spectrometry (ESI-ToF-MS). Upon irradiation with UV light, the copolymers degraded rapidly via intramolecular cyclization into small molecules. After 15 min irradiation, more than 96 % of protecting groups were cleaved, while the number averaged molar mass (Mn) decreased from 4000 g/mol to 500 g/mol as detected by SEC. Afterwards, the degradation products were separated and identified by coupling of advanced polymer chromatography (APC) and ESI-ToF-MS. The synthesized light sensitive biodegradable copolymer was expected to be applied for precisely controlled release.

In the present study, LrPs were used to obtain light-triggered degradation properties of the resulting NP. The manufactured NP were loaded with a model drug and characterized in detail.

# Biobased and biodegradable polycondensates from isohexide building blocks

Jing Wu
Donghua University

Driven by the desire for a sustainable society, development of biobased and biodegradable polymers has great economic and ecological potentials. Isohexides (1, 4;3, 6-dianhydrohexitols) are a group of highly interesting carbohydrate-based building blocks, which are featured by their high structural rigidity, high hydrophilicity, chirality and low toxicity. By taking advantage of these characteristics, isohexides have been widely used in synthesizing various types of polycondensates. This report focus on the biobased and biodegradable polycondensates systems based on isohexides and their derivatives. The synthesis and structure-properties are systematically studied. The effect of isohexides incorporation on the overall thermal, mechanical and (bio)degradabilities will be illustrated.

# Measurement of fabric shear in drape taking into account three-dimensional deformation of grainlines

Liu Yang, KyoungOk Kim, Masayuki Takatera
Shinshu University

As a typical three-dimensional large deformation, fabric drape and its mechanical properties have been investigated since 1960s. In Fabric Research Laboratories (FRL) drape test, drape coefficient (DC) is a wildly used index to evaluate fabric drapability[1]. It is defined by using the ratio of the projection areas of a piece of circular fabric in its flat and draping states. The effects of bending rigidity and weight on bending and drape of fabric are investigated by Peirce, Bickley, Cusick and other researchers [3-5]. Yang et al.[6] analyzed the effects of fabric dimension on drape deformation using a model of a circular segment cantilever for infinite shear stiffness (upper limit) and the deflection of strip cantilevers in radial directions for zero shear stiffness (lower limit). However, finite shear deformation could occur in actual drape on conditions that fabrics have small shear stiffness, large dimension, or heavyweight. To confirm the effect of shear stiffness, they also proposed a measuring method to measure shear deformation using three-dimensional scanning [7]. They assumed that the center grainlines were straight on drape. However, the grainlines deformed three-dimensionally along to drape, which lead to a different calculation in analyzing shear deformation using shear angle. Therefore, in this study, we recalculated the shear deformation using the deformed grainlines.

# Preparation and characterization of carbon fiber reinforced multi-phase epoxy syntactic foam filled with carbon fiber reinforced hollow epoxy macrospheres and hollow glass microspheres

Ying Wang, Yuan Gao, Tao Jiang, Zhongxian Zhao, Yuantao Zhao, Wenge Li, Xinfeng Wu
Shanghai Maritime University

Carbon fiber reinforced hollow epoxy macrospheres (CFR-HEMS), hollow glass microspheres (HGMS), and carbon fiber (CF) were added into epoxy-hardener system to prepare multi-phase epoxy syntactic foam (ESF) using molding method. An innovative “rolling ball method” was implemented in the preparation of CFR-HEMS where expanded polystyrene (EPS) beads were used as initiation template. The EPS beads were coated with the epoxy resin and carbon fiber using “rolling ball method”, and these coated EPS beads were later cured and post-cured at high temperature which will shrink the EPS beads thus producing a hollow macrosphere structure. Scanning electron microscope (SEM) shows that the “rolling ball method” can make carbon fibers form a fiber spherical x-y network throughout the macrosphere wall, which can make CFR-HEMS and multi-phase ESF have great compressive strength. In the process of preparing CFR-HEMS, carbon fiber is added to enhance the strength of the epoxy hollow sphere. In the process of preparing the multiphase ESF, carbon fiber is added to enhance the strength of the matrix epoxy resin and then enhance the multiphase ESF. The carbon fiber reinforced ESF (506 kg/m3, 26.38 MPa) can withstand the 2638 meters water pressure and provide 494 kg/m3 buoyancy, which can give some advice to the preparation of buoyancy material used in deepwater oil exploration.

# Preparation and comprehensive properties of lightweight carbon/carbon fiber composite

Bo Tang[1, 2], Ying Wang[2, 3], Jing Luo[4], Tengfeng He[4], Yufeng Wang[4], Chongyin Zhang[4], Yonggen Lv[1], Xinfeng Wu[2, 3]
1.Donghua University
2.Hangzhou Vulcan New Materials Technology Co., Ltd
3.Shanghai Maritime University
4.Shanghai Aerospace Equipments Manufacturer Co., Ltd, Shanghai Engineering Reseach center of Specialized Polymer materials for Aerospace

In this paper, lightweight carbon/carbon fiber composite thermal field insulation materials were prepared by the process method of carbon fiber airflow netting-needle punching forming felt-resin impregnation-molding curing-high temperature carbonization and graphitization. The graphitization, the microscopic morphology, conductivity, bending strength and thermal conductivity of carbon/carbon fiber composites were measured by X-ray Diffractometer (XRD), scanning electron microscope (SEM), four probes, electronic universal testing machine and thermal analyzer. The main conclusions are as follows: Carbon/carbon fiber composite has been fully graphitized after high temperature treatment. The results show that the long carbon fiber in the carbon/carbon fiber composite forms a three-dimensional structure of X-Y-Z with a density of 0.16±0.02g/cm3, which makes the composite material have excellent thermal insulation performance at high temperature, and the conductivity is 1.45*103 S/m (plane direction) and 18.2*103 S/m (perpendicular to plane direction), the bending strength is 3.52 MPa (plane direction) and 1.86 MPa (perpendicular to plane direction), the thermal conductivity is 0.076 W/(mK) (25 ℃) and 0.17 W/(mK) (1200℃), respectively. The above process methods and test results will provide the application of carbon/carbon fiber composite in solar polysilicon furnace, single crystal silicon furnace, semiconductor furnace, sapphire furnace, fiber drawing furnace, high-end metallurgical heat treatment furnaces, as well as applications in other high-temperature insulation environments and also provide some suggestions in these insulation applications.

# An overview of industrialization development of lightweight carbon/carbon fiber composite for high temperature furnace

Bo Tang[1, 2], Ying Wang[2, 3], Mingliang Chen[4], Tongliang Du[4], Xun Ding[4], Chongyin Zhang[4], Yonggen Lv[1], Xinfeng Wu[2, 3]
1.Donghua University
2.Hangzhou Vulcan New Materials Technology Co., Ltd
3.Shanghai Maritime University
4.Shanghai Aerospace Systems Engineering Institute, Shanghai Aerospace Equipments Manufacturer Co., Ltd, Shanghai Engineering Reseach center of Specialized Polymer materials for Aerospace

Lightweight carbon/carbon fiber composites are used in high temperature equipment such as solar polysilicon furnace, single crystal silicon furnace, semiconductor furnace, sapphire furnace, fiber drawing furnace, high-end metallurgical heat treatment furnaces for their excellent thermal insulation, high temperature resistance, high purity and mechanical properties. In this paper, the product introduction, preparation method, product performance, manufacturer, market application, market capacity and market development direction of light carbon/carbon fiber composites are briefly described. It is hoped that this paper can provide some help for the industrialization of light carbon/carbon fiber composites. At the same time, it is hoped that this can promote the application of canbon/carbon fiber composite in high temperature equipment.

# Synthesis of four-armed star-shaped poly(N, N-diethylacrylamide) by group transfer polymerization in the presence of hydrosilylane

Jian Li, Toyoji Kakuchi, Xiande Shen
Changchun University of Science and Technology

The core-first synthesis of four-armed star-shaped poly(N, N-diethylacrylamide)(s-PDEAM) with predicted molecular weights and narrow molecular weight distributions (Mw/Mn=1.17) was carried out by the Tris(pentafluorophenyl)borane (B(C6F5)3) -catalyzed group transfer polymerization (GTP) using ethyldimethylsilane (EtMe2SiH) as hydrosilylane. The optimal ratioof the precursor to the hydrosilylation agentwas1:4.2. For the s-PDEAMs, the synthesis of s-PDEAM was carried out under the conditions of [DEAM]0/[s-SKA3]0= 200, 400, and 800. All the SEC(RI) traces showed a unimodal distribution and shifted to the higher molecular weight region with the increasing [DEAM]0/[C(MA)4]0ratio.The Mw/Mns values were as narrow as 1.21 - 1.30 and the 1H NMR results mean that the core-firstGTP is a useful method for the synthesis of the star-shaped polyacrylamide.

# A novel waterborne fluorescent nano-fiber membranes with effectively suppressed ACQ phenomenon: fabrication, properties and applications

Weili Li[1], Lin Wang[1], Jun Wang[1], Lei Yang[1], Yixiao Xie[1], Mike Tebyetekerwa[2], Shengyuan Yang[3]
1.Jiangsu University of Science and Technology
2.The Australian National University
3.Donghua University

In this research, a type of waterborne fluorescent nano-fiber membranes(WNFM) based on fluorescent acrylic latex(FAL) was obtained by electrospinning. At first, the fluorescent latex was obtained by doping Rhodamine B (Rh B) into acrylate monomers during emulsion polymerization. Then PVA(poly vinyl alcohol) was added into and a series WNFM were fabricated via electrospinning by adjusting the parameters. Due to electrostatic attraction, a massive Rh B was absorbed and dispersed onto the shell of the latex, which avoided its aggregation caused quench phenomenon(ACQ). The relationship between the preparation parameter, morphology and properties of WNFM were studied systematacially. Especially, due to sensitivity and accuracy, the fluorescence spectrum was applied in exploring the property of the prepared WNFM, which opens a new way to explore the process of electrospinning with water as the spinning solvent. As the fluorescent property of Rh B can be impacted by the surrounding stimuli, the fluorescent responsiveness property of the obtained WNFM was studied to explore its application.

# Fast preparation of high molecular weight polypeptide by ring opening polymerization of N-carboxyanhydrides (NCA) outside of the glovebox

Yueming Wu, Runhui Liu
East China University of Science and Technology

Polypeptides have broad applications such as tissue engineering, drug delivery and antimicrobials. It has been more than 110 years since Hermann Leuchs discovered N-Carboxyanhydrides (NCA) for the first time in 1906 and since then NCA polymerization has developed into a classical method for the preparation of polypeptides. However, NCA polymerization is sensitive to moisture and the most widely used primary amine initiator only gives slow polymerization resulting polypeptides in short chain length. Therefore, we developed the LiHMDS-initiated open vessel NCA polymerization that is not sensitive to moisture and can initiate NCA polymerization outside of the glovebox without any protection. The simple operation facilitates large scale and parallel synthesis of polypeptide. In addition, the polymerization can complete as fast as mins and it is suitable to prepare long chain polypeptides that are important for many biological activities and functions.

# An investigation of the thermal and (bio)degradability of PBS copolyesters based on isosorbide

Jiefei Qi
SAIC DM Co.

As an essential biodegradable polyester species, poly(butylene succinate) (PBS) is restricted for wider applications due to its low thermal/mechanical properties and unsatisfactory (bio)degradability. Current practice of introducing either stiff or flexible building blocks into PBS main chains remains challenging to achieve a synergistic enhancement of the thermal and (bio)degradability of this material. We herewith report a series of PBS copolyesters based on the carbohydrate-derived isosorbide (1, 4:3, 6-dianhydro-D- glucidol, IS) (PBIS) by utilizing its unique intrinsic characters of being rigid and hydrophilic. The target copolyesters were constructed with a broad scope of IS content (0—100 mol%) and with random microstructures. The Mn values and the intrinsic viscosities of these polyesters are in the scopes of 7300 —38, 700 g mol-1 and 0.33—0.82 dL g-1 respectively. The results shown in this work clearly demonstrated that the presence of IS enhances the Tg values almost linearly and simultaneously promotes (neutral, acidic) hydrolytic and enzymatic degradations (with porcine pancreas) of the copolyesters. PBIS copolyester containing 20 mol% IS displays comparable hydrolytic and enzymatic degradation rates with those of PBSA (20 mol% adipic acid), but a substantially 23℃ higher Tg value. Detailed characterization of the molecular structures, micro-sequential structures, molecular weights and polydipersities, thermal properties, hydrophilicities and (bio)degradability are provided. The (bio)degradation and degradation mechanism study of these copolyesters are reported for the first time.

# Preparation and study of high temperature resistant polyphenylene sulfide fiber

Guofeng Yang, Zhenhuan Li, Kunmei Su
Tiangong University

In this study, oxidative crosslinking modification of polyphenylene sulfide fibers was carried out using hydrogen peroxide as oxidant, glacial acetic acid and boric acid as catalysts. High temperature resistant and anti-oxidation polyphenylene sulfide fiber was prepared by adjusting the reactant ratio and reaction temperature.The structure and properties of the fibers were characterized by DSC, mechanical properties tests and high temperature resistance tests.DSC test results show that the crystallinity of the modified polyphenylene sulfide fiber is reduced, and the fiber is transformed from a semi-crystalline structure to an amorphous structure. The mechanical properties test and high temperature resistance test show that the tensile strength of the modified fiber is reduced by 25%, but the high temperature resistance is significantly improved, the tensile strength of the modified fiber can still be maintained at 90% when heated at 350 ° C for 2 h. The final results show that when the mass ratio of polyphenylene sulfide fiber, hydrogen peroxide, glacial acetic acid and boric acid is 2:30:20:1 and reacted at 95 °C for 2 h, the prepared fiber has the most excellent comprehensive properties.

# Improvement of elasticity of bio-based poly(trimethylene terephthalate) fibers via addition of graphene

Jun Sun, Lixing Dai
Soochow University

Poly (trimethylene terephthalate) (PTT) fiber has made great progress after breaking through the limitation of raw material bio-based 1, 3- propylene glycol. As reported, the trans-gauche-gauche-trans conformation in the O-CH2-CH2-CH2-O part of the macromolecular chain of PTT crystals showed repeat units distinct "Z" shape conformation, so that the helical PTT molecular chain was like a spring coil. In addition, the crystal modulus of the helical conformation of PTT molecular chain has a particularly low modulus of only 2.49 GPa[1],which is much smaller than the 105 GPa of the PET[2]. Therefore, the intrinsic excellent elastic recovery of PTT fiber depends on the conformation in the crystal region and the order degree of the aggregation state of the molecular chain in the amorphous region. However, in actual use, the elastic recovery of PTT fibers tends to decrease due to heat or external force. Therefore, how to further strengthen the elastic recovery of the fiber based on the intrinsic elastic recovery of the PTT fiber has already attracted great interest from the relevant researchers.

As we know, graphene(GP) possesses the excellent electrical, thermal and mechanical properties, and recent study showed that GP could increase the elasticity of polymeric materials [3]. In this paper, the composite fiber of PTT/GP was prepared by melt-spinning through die extrusion of a double-cone twin-screw extruder and the structure and elastic recovery properties of the composite fiber were studied. The results showed that there was a narrow and sharp crystallization peak on the DSC curve of PTT/GP fiber, while there was a wide and flat crystallization peak on the DSC curve of PTT fiber, which means the addition of GP obviously promotes the crystallization of PTT. In order to study the elastic recovery of PTT/GP composite fiber, we measured the load–elongation curves of drawn fibers in five stretch cycles by elongation of 30%, the results showed that the elastic fiber recovery rate could be increased from 55.78% of pure PTT fiber to more than 60% of PTT/GP composite fiber. Except for the crystallization promotion of GP to PTT, the stiffness of GP relative to the polymer chains could help the increase of the fiber elasticity.

# Hydraulic permeability of quasi-parallel fibrous membranes by electrospinning

Zeman Liu, Fei Guo
School of Energy and Power Engineering, Dalian University of Technology

Fibrous membranes with nonwoven structure and quasi-parallel fibrous structure are fabricated by electrospinning technique. The quasi-parallel fibrous membranes and nonwoven membranes exhibit different behaviors to the hydraulic flow passing through the membranes. The effects of the fiber arrangement, fiber diameter, and deformations of the fibers on the hydraulic permeability are studied in detail. The results show that the hydraulic flow can generate extrusion pressure which affects the porosity and pore structure of the fibrous membranes. Quasi-parallel fibrous membranes and nonwoven membranes exhibit the similar variation tendency to the change of experimental variables. However, the quasi-parallel fibrous membranes exhibit higher sensibility to the change of the hydraulic flow rate. The hydraulic permeability of quasi-parallel fibrous membranes is further analyzed with packing state models in this work.

# Effect of NaSCN content on the rheological Behavior of PAN/NaSCN aqueous solution

Lingdong Shi[1], Yongbo Yao[2], Yumei Zhang[1]
1.Donghua University
2.Jiaxing University

Polyacrylonitrile (PAN) fibers obtained by dry-jet wet spinning have the advantages of high crystallinity, high orientation, and good mechanical properties [1]. Compared with other methods. PAN solutions that have been industrially used for dry-jet wet spinning are formulated with organic solvents, which have the advantages of strong dissolving ability, high solubility, and can prepare solutions with high viscosity. Compared to organic solvents, although NaSCN aqueous solution has the disadvantage of low solubility, it has the advantages of fast reaction rate of acrylonitrile, small chain transfer constant and high conversion rate [2]. Moreover, no harmful substances are volatilized during the spinning process, and the recycling technology is mature, so it is still the main solvent used in the spinning of the PAN solution. Among the key factors affecting spinnability of dry-jet wet spinning of solution [3], NaSCN content has significant effect on the solubility of PAN and elastic viscosity of its aqueous solution for the PAN/NaSCN aqueous solution.

In order to study the effect of NaSCN content on the stability and viscoelasticity of PAN/NaSCN aqueous solutions, the rheological behavior of PAN/NaSCN aqueous solution with different NaSCN content was studied.

From Fig.1, it is shown that the viscosity and relaxation time of the solution increased with increasing NaSCN content while the viscous flow activation energy changed little with the NaSCN content in the solution between 44.1wt% and 49.3wt%. Dynamic frequency measurement has proved that modulus index is unchanged (shown in Fig.2), which indicates that the solutions are homogeneous and stable.

Corresponding to the rules suggested by Winter et al[4, 5], the frequency independence of loss tangent tan δ indicates that there is no gelation transition during cooling as shown in Fig.3, which indicates that the solution is stable under the experiment condition.
The present work takes a further study of the rheological of PAN/NaSCN aqueous solution, which provided a reference for the design of spinning for the design of a solution suitable for dry-jet wet spinning.

**POSTER PRESENTATION**

# Antibacterial and anti-fibrillation modification of lyocell fibers

Tiehan Wang[1], Wei Yuan[1], Yongbo Yao[2], Yumei Zhang[1]
1.Donghua University
2.Jiaxing University

Cellulose, as one of the most abundant natural resource is widely used in textiles, food, medical care, nursing, packaging and other fields. However, since cellulose has a hydrophilic structure, it is an advantageous medium for the reproduction of microorganisms. Therefore, the development of antibacterial cellulose products has been attracted much attention [1] [2]. In order to develop a kind of antibacterial and environmental friendly regenerated cellulose fiber, the low crystallinity and loose structure of the primary gel state Lyocell fibers were selected as the based materials. The quaternary ammonium salt-containing antibacterial agent containing methoxysilane was subjected to a siloxane coupling condensation reaction with the hydroxyl group of the cellulose to carry out antibacterial modification while reducing the fibrillation tendency of the fiber, shown in Fig.1.

KEYNOTE SPEECH

# Two-dimensional covalent organic frameworks with hierarchical porosities

Xin Zhao
Shanghai Institute of Organic Chemistry, Chinese Academy of Sciences

Porous structures with hierarchical porosities are widely found in Nature and play crucial roles in implementing biological functions. Studies on artificial porous materials have also revealed that hierarchical porous materials display advantageous properties such as improved mass transport, increased distribution of active sites, and minimized diffusion barriers. Such advantages make them highly attractive for catalysis, energy storage and conversion, sensing, and separation. As a class of porous organic polymers with precise molecular design, tunable structures and high crystalline networks, covalent organic frameworks (COFs) are an excellent platform for the fabrication of hierarchically structured porous materials. In 2014 we reported the first example of two-dimensional (2D) COF possessing two different kinds of pores, in which trigonal micropores and hexagonal mesopores are periodically integrated. This work, together with a series of succeeding research, has developed a new type of COFs termed heteropore COFs which exhibit hierarchical/heterogeneous porosities. The emergence of heteropore COFs not only increases diversity of COFs, but also brings in new properties and applications, as a result of their highly ordered hierarchically/heterogeneously porous structures. In this presentation, this emerging class of crystalline porous materials will be reported.

INVITED LECTURE

# 2D polymers: next generation of polymers

Yingjie Zhao
Qingdao University of Science and Technology

Highly conjugated two- and three-dimensional (2D and 3D) covalent organic frameworks (COFs) were constructed by delicate molecular design and different chemistry. A triazine based fully conjugated 2D COF bridged by -C=C- linkers was synthesized. This 2D COF shows good crystallinity and excellent solid luminescence property. The application of coenzyme regeneration and coenzyme assisted synthesis of L-glutamate was explored. This fully conjugated 2D COF shows unprecedented coenzyme regeneration efficiency and record-breaking results of the coenzyme assisted synthesis of L-glutamate. In addition, highly conjugated 3D COFs were constructed based on spirobifluorene cores linked via imine bonds with novel interlacing conjugation systems. The crystalline structures were confirmed by powder X-ray diffraction and detailed structural simulation. A 6- or 7-fold interpenetration was formed depending on structure of linking units. The obtained 3D-COF showed permanent porosity and high thermal stability. In application for solar cells, simple bulk doping of 3D-COF to the perovskite solar cells (PSCs) substantially improved the average power conversion efficiency (PCE) by 18.0%.

**ORAL PRESENTATION**

# Morphology design of covalent organic frameworks

Song Wang, Pingwei Liu, Wen-Jun Wang
Zhejiang University

Covalent organic frameworks (COFs) possessing well-defined nanopore structures can precisely control the transport of molecules through them, thereby promising applications in separations, gas storage, sensing, and catalysis. However, such applications are compromised by the low crystallinity and thus poor morphology control of COFs. Herein we introduce a Reversible Polycondensation-Termination (RPT) approach. By simultaneously introducing two monofunctional competitors into reaction systems, the RPT approach selectively generates spherical, fibrous, and membranous COFs with highly ordered structures up to centimeter dimensions. The monofunctional competitors can reversibly terminate/activate the polycondensation reaction between monomers during the COF synthesis by dynamically combining with reactive functional groups of monomers. As a proof of concept, we applied the COF film for microreactions with high catalytic activities and also rapid vapor sensor with repeatable color change.

KEYNOTE SPEECH

# Semiconducting 2D vinylene-linked covalent organic framework nanofibers

Fan Zhang, Shuai Bi
Shanghai Jiao Tong University

Construction of organic semiconducting materials with in-plane π-conjugated structures and robustness through carbon-carbon bond linkages, alternatively as organic graphene analogs, is extremely desired for powerfully optoelectrical conversion, energy conversion and storage. However, the poor reversibility for sp2 carbon bond forming reactions makes them unavailable for building high crystalline well-defined organic structures through a self-healing process, such as covalent organic frameworks (COFs). Here, we are presenting a scalable solution-processing approach to synthesize two-dimensional (2D) COFs with vinylene linkages by using electron-deficient aromatic compounds as the key building blocks to promote an organic base-catalyzed Knoevenagel condensation with various aldehyde-substituted arenes. The resulting carbon-carbon double linked frameworks, feature crystalline honeycomb-like structures with high surface areas and substantial open regular channels. These COFs shows uniform nanofibrous morphologies, allowing for further assembled with carbon nanotubes to form the flexible thin films. Such films are facile to be fabricated to interdigital electrodes used for a microsupercapacitor (MSC). These COF-based MSC devices exhibit excellent electrochemical performance with high power and energy densities, as well as excellent flexibility, and integration capability, suitable as the power source for wearable/portable systems.

INVITED LECTURE

# Design of porous organic polymers with structure-enhanced photothermal conversion performance

Jia Guo
Fudan University

Exploration of the underlying relationship between topological structure of porous organic polymers (POPs) and their performances has long been at the heart of material application in the subject. Photothermal conversion, referring to transduction of light energy into heat, has been developed for the treatment of various medical conditions. This technique is an extension of photodynamic therapy, in which a photosensitizer is exited with specific band light and release energy in term of heat. Of photosensitizer varieties, inorganic nanomaterials such as gold, graphene, carbon nanotube and so on, have exhibited outstanding performances in photothermal conversion upon irradiation of near-infrared (NIR) light. On the other hand, taking in vivo application into account (e.g. compatibility, cytotoxicity, degradation, metabolism, etc.), researches on organic photosensitizers have remained active as well in view of distinguished photophysical properties and bioavailability of organic dye molecules. However, development of polymer-based photosensitizers has lagged behind the others in this area, although they have usually been regarded as a versatile and flexible platform enabling multi-function integration.

In the context, we have comprehensively studied the photothermal performances of POPs by modulating topological structures, which are typically composed of covalent, reticular, and porous skeletons within such as 2D covalent organic frameworks (COFs) and 3D conjugated microporous polymers (CMPs). As illustrated here, our findings have demonstrated the essential role of molecular arrangement within specific motifs both on NIR absorbability and NIR-triggered photothermal conversion. Briefly, given the CMPs containing strained macrocycles in closed networks, topological polymerization of dye molecules could modulate the π-electronic interaction to implement efficient light-to-heat transduction. Analogously, modular design of COFs not only promotes the π-π stacking energy of atomic layers, but also facilitates generation and delocalization of radicals within frameworks, both of which result in the red-shifted and intense absorption and fluorescence quenching, thereby releasing excited energy in term of heat. In the estimation of photothermal efficiency, the topologically designed POPs outperform the linear and branched counterparts, and have been ranked among the best of organic photosensitizers. Our studies shed light on an innovative route to design polymer photosensitizers with prominent photophysical properties based on topological chemistry of POPs.

INVITED LECTURE

# Electro-optically switchable conjugated microporous polymers for environmental and green energy applications

Yaozu Liao
Donghua University

Development of electro-optically switchable conjugated microporous polymers (CMPs) combing π-conjugation of the organic backbone with permanent micropores, has received increasing attention for many applications such as energy storage and conversion, photocatalysis, and chemosensors. These new materials are usually formed by covalent bond linking with rigid building blocks having a geometric shape via molecular design methods. In order to obtain electro-optically switchable CMPs with specific structures and regulatable properties, it is important to study its topology structure and to objectively select suitable structural units and efficient polymerization. Nitrogen-enriched high surface area CMPs are in particular interesting owing to their polarity, basicity, and coordination ability to metal ions. In this presentation, several efficient polymerization coupling methods such as Buchwald-Hartwig, Chichibabin, and Schiff-base reactions will be introduced to synthesize electro-optically switchable CMPs. It is found that careful selection of the types and topologies building blocks, as well as polymerization methods, enables the surface areas, pore sizes, optical band gaps, and electrochemical activities to be controlled. The relationship between the molecular design of such CMPs and their versatile applications including capture and storage of greenhouse gas carbon dioxide and radioactive iodine, supercapacitor energy storage, photocatalytic and electrochemical hydrogen evolution will be discussed.

ORAL PRESENTATION

# Flexible fiber-shaped supercapacitors with excellent energy density based on carbon nanotubes/conjugated microporous polymer networks

Wei Lyu, Yaozu Liao
Donghua University

Flexible fiber-shaped supercapacitors have recently attracted considerable interest due to their small size, light weight and potential application in wearable electronics and smart textiles. However, the low energy density makes them slightly inferior than batteries, which limits their practical application. Conjugated microporous polymer (CMP) networks featuring designable porous structures and high permanent surface areas have a huge potential for high-performance energy storage owing to the rapid ion and electron transportation. Here, poly(triphenylamine) (PTPA), a N-containing CMP, was in-situ synthesized on the p-bromobrnzene-modified carbon tube fibers (CNTF@PTPA) by Buchwald-Hartwing cross-coupling reactions. An all-solid-state twisted symmetric fiber-shaped supercapacitor was assembled by wrapping the gel electrolyte. The resulted supercapacitors exhibited a broadened potential window of 1.4 V, a high areal specific capacitance of 105.67 mF cm-2 and a maximum energy density of 21.5 μW h cm-2, which is much higher than other reported fiber-shaped supercapacitors. Meanwhile, the all-solid-state CNTF@PTPA showed excellent mechanical stability, with 84.5% of the initial capacitance can be retained after 10000 knotting cycles.

**KEYNOTE SPEECH**

# 1D materials: sensors and energy storages

Kim, Il-Doo*

Korea Advanced Institute of Science and Technology

Electrospinning has been recognized as one of the most efficient techniques for producing non-woven fiber webs on the order of several hundreds of nanometers by electrically charging a suspended droplet of polymer solution with/without inorganic precursors or melt. Various types of materials with a high degree of porosity, a large surface area, superior mechanical properties and modified surface functionalities, can be electrospun into nanofiber structures. These materials include polymeric nanofibers as well as metallic and metal-oxide nanofibers which are prepared by a subsequent heat treatment in a reducing or oxidizing atmosphere of metal salt precursor/polymer composite fibers.[1] In particular, the simplicity of the process combined with the possibility of large-scale production through the use of multiple-nozzles (> 10, 000 pieces) makes this process very attractive and therefore opens up new commercial markets for diverse applications. In this presentation, I summarize recent progress and a collection of advances, particularly focused on the synthesis, characterization, and utilization of electrospun metal oxide nanofibers. I will end my presentation by suggesting possible future research direction and potential suitability of 3D nanofibers for applications in colorimetric sensors, exhaled breath gas analyzing sensors for early stage disease diagnosis as well as energy storage applications including lithium or zinc ion batteries.

INVITED LECTURE

# Functionalization of metal-organic frameworks for broadband white-light emission and chemical $CO_2$ fixation

Honghan Fei
Tongji University

Organolead halide perovskites are an emerging class of intrinsic broadband white-light emitters with high color rendering, high tunability and facile processibility, but suffer from a variety of limitations including low quantum efficiency, low moisture stability and difficulty to functionalize. The presentation will introduce our efforts to employ anionic structure-directing agents to template a family of 2D cationic lead halide layered frameworks, occupying high-coordinate halide ligands and ultrahigh chemical stability[1-3]. We continue to extend the research to discover metal-organic frameworks (MOFs) consisting of structurally deformable metal halide secondary building units (SBUs)[4]. In contrast to organolead halide perovskites, MOFs with one-dimensional lead halide units have advantages such as high moisture stability, facile functionalization and incorporation of luminescent guest molecules.

The second part of the presentation will change the topic from the functionalization of SBUs in MOFs to functionalization of organic ligands in MOFs, focusing on the catalytic applications in chemical CO2 fixation. One example is to functionalize UiO-68 with N-heterocyclic carbene (NHC) moieties for highly efficient and selective hydrosilylation of CO2[5]. Postsynthetic ligand exchange is an important approach to achieve in-situ deprotonation of imidazolium salts during postsynthetic functionalization, achieving one rare MOF example with metal-free NHC species. Computational studies indicate the confined MOF porosity provides an excellent platform for activation of Si-H bonds and high selectivity toward high-value silyl methoxide product, compared to the homogeneous NHC analog. Our recent studies on organosulfonate-based MOFs for CO2 capture and conversion will be discussed as well[6, 7].

INVITED LECTURE

# Archimedean spiral fibers with continuous two-dimensional inclusions

Pingwei Liu
Zhejiang University

Graphene and other two-dimensional (2D) materials are distinct among nanoscale inclusions or fillers for composites in that they can potentially span the physical dimensions of the enclosing solid. However, alignment and assembly of continuous 2D components at high volume fraction and macroscopic dimensions remains an unsolved challenge in material science. Herein, we introduce a transverse shear scrolling method that generates Archimedean spiral nanocomposite fibers 0.10-0.16 mm in diameter and 2 cm in length. The process significantly increases the effective elastic modulus approximately 1.9-fold to nearly 1 GPa, approaching the limit of platelet filler theory, and increases the ultimate tensile strength to 40 MPa even at an exceptionally low graphene volume fraction of only 0.00185. The fibers demonstrate exotic, telescoping elongation at break of 110%, or 30 times greater than Kevlar. Both composite types retain anisotropic electrical conduction along the graphene (or $MoS_2$) planar axis with a percolation threshold VG < 0.003 vol%, and layer numbers less than 36 remain transparent with optical density < 42%. We also demonstrated that the fibers could be used as optical sensors to monitor the strain and temperature if $MoS_2$ was used. These results highlight new combinations of material properties available at this extreme platelet filler limit for fiber nanocomposites.

**INVITED LECTURE**

# Nanofibres for the environment: transitioning toward sustainability

Addie Bahi[1], Frank Ko[2]
1.Research Scientist
2.Professor

After gaining his Ph.D. degree in Biomaterials, Addie joined AFML at UBC in 2010. His multidisciplinary educational background in Fibre Science, Chemical Engineering, Biotechnology, Software Systems, Management, and Materials Engineering assists him in exploring integrated engineering and interdisciplinary collaboration with academia and industry partners. Dr. Bahi has maintained multiple collaborations with Professors in Forest Sciences, Civil Engineering, Microbiology & Immunology, the Beedie School of Business, Chemical and Biological Engineering, Mechanical Engineering, Electrical and Computer Engineering, and Agriculture. On the industrial side, these collaborations resulted in collaborative proposals and projects with more than 30 industry partners.

Human impacts on the environment raise consciousness requiring not only serious and immediate attention but also action. Considering the limited resources and materials available, circular economy principles can provide a practical platform for an appropriate response, where one industry's waste becomes another's raw material.

Lignocellulosic materials (LCMs) are the most abundant materials on the planet. Converting LCMs into nanofibre can create the potential for manufacturing value-added products. This work will provide and expand two examples in this field.

**KEYNOTE SPEECH**

# Nanofibrous energy storage materials and laser ultrasound transducers

Xiangwu Zhang
Wilson College of Textiles, North Carolina State University, Raleigh, North Carolina

Nanofibers are an important class of material that is useful in a variety of applications, including filtration, tissue engineering, protective clothing, composites, battery separators, energy storage, etc. So far, electrospinning is the most studied method for producing nanofibers. However, the wide-spread commercial use of electrospinning is limited due to its low production rate, poor safety, and high cost. Most other nanofiber production methods, such as phase separation, template synthesis, and self-assembly, are complex and can only be used to make nanofibers from limited types of materials. This presentation introduces a simple, yet versatile technique for producing nanofibers of various materials including polymers, carbons, ceramics, metals, and composites. Centrifugal spinning eliminates the limitations encountered by current nanofiber production methods and can produce nanofibers at high speed and low cost. Centrifugally-spun nanofibers can be used in various applications, including filtration, tissue engineering, protective clothing, composites, battery separators, energy storage, etc. The preparation and applications of centrifugally-spun nanofibers as energy storage materials and laser ultrasound transducers will be discussed in this presentation.

**INVITED LECTURE**

# Functional fibrous structures for mechanical energy harvesting

Jian Fang*[1], Yuying Cao[2]
1.Soochow University, China
2.Deakin Univeristy, Australia

Harvesting renewable mechanical energies into electric power through piezoelectric or triboelectric effects is an emerging technology in materials and technology fields. It offers a promising solution to powering microelectronic devices, development of self-powered sensor, robot systems and personal electronics. Compared with traditional planar and rigid devices, power generating devices made from fibrous materials have attracted growing attention owing to their flexibility, breathability and good compatibility with textiles. These attributes offer them outstanding potential in the development of next-generation self-powered wearable electronics. This talk covers our recent contributions to the development of nanofibrous piezoelectric energy harvesters and textile-based triboelectric energy harvesters.

**INVITED LECTURE**

# Bioinspired smart materials with multiscale architecture and multifunctionality

Hao Bai
Zhejiang University

Development of human society is, to some extent, relying on the invention of new materials. In this context, biological materials, such as bone, shell and bamboo, constantly serve as a source of inspiration to design strong, tough, lightweight, self-healing and smart synthetic materials for future engineering applications. While biological materials achieve multifunctionality by building sophisticated multiscale architecture, synthetic materials are always relying on the diversity of constituents. Combining these two strategies would stimulate green fabrication approaches and result in multifunctional materials with unprecedented properties. Specifically, we take an ice-templating technique to mimic the sophisticated architecture of biological materials in our synthetic counterparts. In this talk, I will show the potential of this biomimetic approach, with our recent progress in thermoregulating textiles inspired by polar bear hair, intrinsically self-healable nacre-mimetic composites and so on.

KEYNOTE SPEECH

# Multiscale nanofibers for liquid separation and catalysis application

Yong Zhao, Nü Wang
Beihang University

Controlling liquid transportant through a porous nanofibrous membrane with superwetting property is of great interests for separation of liquid mixture, such as free oil/water mixture, emulsion and organic liquid mixture, is widely required in industry process and environmental protection, but how to separate liquids mixture in a high-efficient, low energy mode is still a challenge. We fabricated a series of superwetting liquid separation membranes with special surface structures as well as proper porosity. We fabricated a branch-like nanofibrous membranes. The membrane could separate various neutral and corrosive emulsions in a highly efficient manner. Besides oil and water system, more complex organic liquid mixtures separation is of vital importance that not only prevents the secondary pollution, but also enhances the recycling of the organic liquids. We fabricated wettability-tunable nanofibrous membrane composited of high performance fluoro-polymer as matrix and fluorosilane as surface energy regulator, which can be applied in immiscible organic liquid mixture separation. These membranes are expected to become competitive candidate for complex chemical products separation, resource recycling, environmental protection and heterogeneous catalysis.

INVITED LECTURE

# Heterogeneous scorpionate site in MOF: Small molecule binding and activation

Le Wang
Donghua University

Scorpionate ligands are widely used in coordination chemistry and catalysis. These ligands are usually neutral or mono-anionic, hence the scorpionate transition metal complexes are often bonded to counter anions that blocking catalytic sites. The metal-organic framework reported here, CPF-5, presents a unique di-anionic scorpionate backbone, therefore allows multi-guest molecule binding.

We have demonstrated the CPF-5 scorpionate site can be utilized as a multi-guest binding site for various Lewis bases. CPF-5 could act as a crystalline sponge for single-crystal X-ray structural characterization of a variety of compounds using coordinative alignment, second coordination sphere interactions, or both.
Moreover, the C-H bonds in these binding small molecules could be aminated by the CPF-5 scorpionate catalytic sites in the presence of nitrene precursors. This Mn-based outer-sphere catalyst promotes the direct amination of C–H bonds in an intermolecular fashion with unrivaled activity producing >105 turnovers. And CPF-5 could be even recycled and reused, it is found to be the only living catalyst for C–H activation to date.

INVITED LECTURE

# Strain-insensitive biomimetic fiber material with high elastic conductivity

Mingwei Tian[1], Tailin Xu[4], Xuqing Liu[3], Xueji Zhang[2], Lijun Qu[1]
1.Qingdao University
2.Shenzhen University
3.Manchester University
4.Beijing University of Technology

Elastic conductive fiber is the basis of flexible intelligent wearable device with broad application prospects. At present, the most widely reported strain sensor with variable resistance and tensile change is that people pursue high flexibility and high sensitivity while ignoring the stable transmission output of electrical signals. The importance of it. Inspired by the changes in body structure when worms crawl in nature, Professor Qu Lijun's research group constructed superelastic conductive fibers with worm-like graphene conductive microlayers on the surface of polyurethane elastic composite filaments by "prestressing method". The fiber has a reversible electrical signal response in the stretch ratio of 815%, wherein the conductive fiber exhibits an ideal electrical signal response in a 220% draw ratio, which can satisfactorily satisfy the stable transmission of the electrical signal in the complex deformation of the fabric.

**INVITED LECTURE**

# Conformal metal-organic framework coatings on nanofibers for ultra-fast detoxification of nerve agents

Junjie Zhao
Zhejiang University

Nerve agents are highly toxic compounds that can cause serve injury and death in small doses. Terrorist attacks using Sarin and other chemical warfare agents (CWAs) reported in recent years motivate the innovation of new materials for long-term protection in a reduced burden for soldiers and first responders. Metal-organic frameworks (MOFs) are promising catalysts for degrading CWAs, but it is challenging to integrate MOFs into functional filter media and protective garments. In our lab, we developed a series of MOF-based nanofibers capable of detoxifying nerve agents in minutes. We applied ALD $TiO_2$ coatings as nucleation layers for conformal Zr-based MOF thin films on PA-6 electrospun nanofibers. We found that these MOF crystals nucleate and grow directly on and around the nanofibers, indicating strong adhesion to the substrates. We further evaluated the catalytic activities of the MOF-functionalized nanofibers for degrading a simulant Dimethyl 4-Nitrophenyl Phosphate (DMNP) and a nerve agent O-Pinacolyl Methylphosphonofluoridate (GD). Half-lives of DMNP were found less than 8 min with UiO-66-$NH_2$ and UiO-67 coated PA-6@$TiO_2$ nanofibers, while all the Zr-based MOF functionalized nanofibers enables ultra-fast destruction of GD with half-lives as short as 2 min. The results clearly demonstrate the excellent detoxification ability of our MOF-functionalized nanofibers, and show great promise for next generation of gas filters, chemical sensors, and smart textile materials.

INVITED LECTURE

# Fibrous controllable liquid transfer: towards high-performance thin-film devices

Huan Liu
Beihang University

In nature, various biological fibrous systems exhibit unique dynamic wetting properties, which has shown many advantages in inhabiting local environments. However, controllable liquid transfer by an open fibrous system is still poorly understood and remains a challenge, because capillary coalescence is frequently encountered when fiber array interacted with a liquid. Here, we revealed the fundamental of the Chinese brush for its capability in controllable liquid transfer: the unique anisotropic multi-scale structure of the freshly emergent hairs. Drawing inspirations, we developed model devices with flexible conical fibers that allows for direct writing functional micro-lines with 10 μm resolution and nano-thin films, with well-defined profile and uniform distribution on diverse substrates. To be noticed, the fibers-guided directional liquid transfer enables fine controlling the liquid/solid/gas tri-phase contact line under multiple directional stresses. Taking advantages, highly oriented polymer thin film and aligned AgNWs film were fabricated in large scale, based on which high performance of polymer TFTs devices and the anisotropic flexible conductive electrode were developed, respectively. We also demonstrated that the conical fiber array enables direct preparing ultra-smooth QD micro-patterns, and thus a high-performance QLED devices was allowed. We envision that the controllable liquid transfer guided by the conical fibers will shed light on the novel template-free printing of functional photoelectric devices.

KEYNOTE SPEECH

# Thin and small for the future nanolayers and nanoparticles-synthetic routes and applications

Adler, Hans-Juergen P. *[1], Zhu, Mingfang[2,1]
1.Technical University of Dresden
2.Donghua University

The lecture will give an overview about our research fields, which were related to the cooperation with the Donghua University during the last 20 years.

The Nanotechnology will be more and more a substantial part of coatings. Due to new synthetic routes and new high sophisticated analytical methods it is possible to watch nano and molecular structures alone and also in a complicate mixture as matrix, e.g. in coatings. With nanoparticles in a top layer increased properties of scratch resistance or weathering stability will be expected.

Our research has been focused on the interface between a metal as substrate and its interaction with the next organic layer.

1. Metal surfaces are protected by inhomogeneous oxides, which are not stable against external influences. Due to their instability against corrosion a protection layer, mostly an organic coating, is necessary. In spite of all developed technologies corrosion attacks can take place due to failures at the metal/lacquer interface. We have synthesized different reactive molecule forming nanolayers in a self-assembling process and reactive nanoparticles for modification the metal/coating interface to improve the adhesion and to introduce functional materials which can actively participate in corrosion inhibition process.

2. Polymeric particles can be also effectively used for the preparation of the composite materials on sub-micrometer scale. This can be achieved by the introduction of different materials such as conjugated polymers, proteins, semiconductors, metals or metal oxides, bio-minerals in form of nanoparticles into the porous microgel structure or onto the surface of compact latex particles. In this way one can expect formation of multi-functional colloids where typical features of polymeric particles (defined size and morphology, large surface area etc.) can be combined with the properties of incorporated functional materials such as conductivity, magnetic response, catalytic activity etc. This approach can open principally new possibilities for application of colloidal particles in different technological systems.

Controlled incorporation of conducting polymers on the PS/PEGMA particle surface led to the formation of colloidally stable particles with high conductivity and extremely large surface area. These composites have been successfully used for the preparation of the humidity sensors, anti-corrosive coatings, separation of organic molecules, design of electronic devises.

**INVITED LECTURE**

# Applications of nanofibers-based composite scaffolds in bone defects

Hongbing Deng*, Gu Cheng, Xiaowen Shi, Yumin Du
Wuhan University

Bone tissue engineering has been one of the hotspots to treat bone defects. In recent years, various nanofibers-based scaffolds were developed to promote bone healing and regeneration. In case 1, SF/PCL nanofibrous mats were designed to mimic release kinetics in the normal microenvironment with a sustained release of BMP2 and a rapid release of CTGF by the co-axial electrospinning and LBL techniques. These results suggested that BMP2 was successfully incorporated into the core of the nanofibers, and CTGF was attached to the surface of the co-axial fibers to promote angiogenesis in the early period of bone healing by stimulating the expression of VEGF. In case 2, a SF-based scaffold that reinforced by SF-chitin hybrid nanofibers was prepared to imitate cartilaginous ECM and repair the cartilaginous defects. The results demonstrated that the nanofibers and TGF-β1 functionalized scaffold promoted adhesion and proliferation of chondrocytes in vitro, and cartilage regeneration in vivo, thus offering a suitable strategy for the repair of cartilage defects.

INVITED LECTURE

# Synthesis and application of yolk-shell structural porous polymer nanoreactors

Kun Huang, Yang Xu, Buyin Shi, Haitao Yu
East China Normal University

Yolk-shell structural porous nanomaterials with hollow void space represent a unique class of complex hybrids, which have received tremendous attention due to their specific physical and chemical properties. Especially, the void space can provide a unique microenvironment to carry or store a large variety of substances such as metal nanoparticls (NPs). Meanwhile, the permeable porous shell can prevent the undesirable coalescence/aggregation or corrosion/dissolution of encapsulated metal (NPs). Here, we report a synthesis of metal NPs encapsulated in the hollow porous polymeric nanosphere frameworks (HPPNFs) by a combination of hyper-cross-linking mediated self-assembly strategy and ship-in-bottle technology. The resultant Metal@HPPNFs composite nanoreactors exhibited the excellent catalysis activity for different substrates and efficient removal effectiveness towards various organic dyes.

**INVITED LECTURE**

# Interfacial design towards MOF@Polymer and MOF@MOF porous composite materials

Tao Li, Hongliang Wang, Sanfeng He
Shanghai Tech University

Polymers and metal-organic frameworks represent two classes of most chemically and structurally diverse materials. Combining them in a hierarchical order will lead to a library of new functional composite materials with emerging properties. However, the key to the rational design of these materials is to understand their interface. This talk will focus on our recent efforts on understanding MOF/polymer and MOF/MOF interfaces through tuning thermodynamic and kinetic parameters. These efforts led to the development of generalizable methods for the construction of a range of MOF@polymer and MOF@MOF porous composite materials and subsequently a range of unique material properties.

**ORAL PRESENTATION**

# Study of PS filament morphology, mechanical properties and oil sorption application

Abdul Wahid, Zeng Yongchun
Key laboratory of Textile Science & Technology, Ministry of Education.

Polystyrene (PS) fibers have been of increasing interest due to their unique structure and properties. The porous structure of electropun PS filament can make it a promising candidate for high-capacity oil sorbent. The continuous length of PS filament with porous structure produce through conventional electrospinning is reported in this study. PS was chosen as a functional polymer in this process, in which PS tends to produce porous structure on the surface and cross-section of filaments. The morphology and structure of electrospun filaments were observed with the help of Scanning Electron Microscope (SEM) test. The mechanical behaviors of the PS polymers in the continuous length of filament with porous structure were studied. The processing parameters i.e. concentrations of polymer, DMF/THF as a solvents with different amount of LiCl to produce continuous filament with porous structures, and applied voltage are influenced the filament morphology were investigated. The oil sorption experiments show that the PS filament sorbents have the advantage of showing large sorption capacity. It is expected that the filament mats can make a promising candidate for potential applications in those fields such as filtration, separation, and sorption.

KEYNOTE SPEECH

# Pasteur's crystallization revisited: a novel chiral separation strategy by mimicking magnetic beneficiation

Xinhua Wan*, Xichong Ye, Bowen Li, Jie Zhang, Fei Zhang
Peking University

Chiral resolution through fractional crystallization is one of the most economical and convenient methods to obtain large-scale optically chiral compounds for modern pharmaceutics. Although significant development has been achieved since Pasteur separated R- and S-enantiomers of sodium ammonium tartrate in 1848, this method is still fundamentally low efficient, due to low transformation ratio or high labor in separation. In this work, we develop a novel stereoseparation strategy by mimicking magnetic beneficiation for quantitatively separating the crystals of conglomerates. The key of this success is based on a kind of delicately designed magnetic nano-splitters, formed by hybrid assembly of chiral amphiphilic block copolymers with magnetic $Fe_3O_4$ nano particles. The magnetic nano-splitters would be selectively wrapped into the S-crystals, leading to the formation of the crystals with different physical properties from that of R-crystals. As a result of efficient separation under magnetic field, high purity chiral compounds (99.2 ee% for R-crystals and 95.0 ee% for S-crystals) can be obtained in a simple one-step crystallization process, with an extremely high separation yield (95.1%). Moreover, the nano-splitters show expandability and excellent recyclability. Although this strategy only works in the case of conglomerates at this stage, we still envision that our strategy opens a new window to develop novel additives for different scale chiral resolution.

**INVITED LECTURE**

# Nano-voided membranes embedded with hollow zwitterionic nanocapsules for superior desalination performance

Zhijuan Sun, Qian Wu, Changhuai Ye, Wei Wang, Liuchun Zheng, Fengkai Dong , Zhuan Yi, Lixin Xue, Congjie Gao
Zhejiang University of Technology

To lower the initial investment and operational cost of desalination and waste water treatment processes, nanofiltration (NF) membranes need to have high water permeation, ionic rejection and antifouling capability. Recently, complicated Turing-type reaction conditions [Science 2018, 360, 518-521] and sacrificed metal organic frame(MOF) nanoparticles [Nat. Commun. 2018, 9, 2004] have been reported to introduce nano-voids into thin-film composite (TFC) poly-amide NF membranes for better performance. To simplify the process, herein, we report the fabrication of thin-film nanocomposite membranes (TFNM) with controllable nano-voids in the polyamide (PA) layer by including hollow zwitterionic nanocapsules during interfacial polymerization. It is found that embedding hollow zwitterionic nanocapsules (HZNCs) in PA layer could increase the membrane internal free volumes, external surface areas and hydrophilicity, thus enhancing the water permeation and fouling resistant without trading off the rejections for multi-valent ions. For example, water flux of the NF membranes embedded with about 19.0 wt% of HZNCs is greatly increased to 1.7 times (73 L m-2 h-1) of the value of the control TFC NF membrane without HZNCs (43 L m-2 h-1) maintaining 95 % rejection to Na2SO4. Moreover, the effect of HZNCs loadings on the membrane performance was also investigated. This work provided a direct and simple avenue to fabricate advanced desalination membranes with superior separation performance.

INVITED LECTURE

# Reversibly photo-modulating mechanical performance of engineered protein fibers

Kai Liu
Chinese Academy of Sciences Changchun Yingyong Chemistry Institute

Light-responsive materials have been extensively studied due to their high spatiotemporal control and non-invasive characters. Among them, designing new materials with photo-sensitive mechanical behaviors is promising in solid-state photonic switches, optical interconnects, artificial muscles, and chemical sensors. However, it remains a challenge to modulate the stiffness and toughness of bulk material in a reversible way. Here, we demonstrate one new type of bioengineered protein fibers by employing electrostatic interactions between supercharged polypeptides and azobenzene based surfactants. Photoisomerization of the Azo moieties from the E- to Z-form alters the tensile strength, stiffness, and toughness of the bulk protein fibers reversibly. Especially, the increased cation - π interactions of the uncomplexed lysine moieties in proteins and the phenyl groups in Z-form Azo leads to a ~ 2-fold increase in the fiber's mechanics. The outstanding mechanical properties open a pathway towards the development of protein fibers as stimuli-responsive bracing biomaterials.

INVITED LECTURE

# Hierarchical nanostructures via an initiated supramolecular polymerization method

Xiaoyu Li
Beijing Institute of Technology

The construction of hierarchical nanostructures with precise morphological and dimensional control has been one of the ultimate goals of contemporary materials science and chemistry, and the emulation of tailor-made nanoscale superstructures realized in the nature, using artificial building blocks, poses outstanding challenges.

Herein, we report a one-pot strategy to precisely synthesize hierarchical nanostructures through an in-situ initiation-growth process from a liquid crystalline block copolymer, via a supramolecular polymerization approach. All these start from a liquid crystalline (LC) block copolymer, which can form monodisperse cylindrical micelles with a LC core via a self-seeding process. We discovered that by adding a small portion of small molecules as initiators, which can form hydrogen bonds, halogen bonds, quaternization, and coordination complexes with the corona-forming block, uniform cylindrical micelles could be produced in a one-pot manner. By adjusting the assembly conditions, the length of cylindrical micelles can be finely tuned from several hundred nanometers up to several microns with very narrow PDI (<1.03). The formation of supramolecular polymers was further demonstrated to be reversible, due to their non-covalent nature.

More interestingly, not only small molecules but also macromolecules, or nano-objects can also be employed to initiate the assembly to form branched, segmented, hairy plate-like, or star-like supramolecular polymers with hierarchical architectures. The length of these cylindrical arms could be adjusted by tuning the ratio between polymer and these large initiators as well. Meanwhile, the geometry and shape of initiators largely dictated the directions of growing micelles, while the number of cylindrical arms were mostly determined by the dimensions of the initiators.

KEYNOTE SPEECH

# Recent progresses in new functions of advanced nanocomposite gel

Kazutoshi Haraguchi
Nihon University

Stimuli-responsive hydrogels, such as poly(N-isopropylacrylamide) (PNIPA) hydrogel, have been extensively studied from both scientific and industrial application points of view. However, conventional PNIPA hydrogels consisting of chemically crosslinked network had some serious disadvantages, particularly weak and brittle mechanical properties and low stimuli-responsiveness, because of the network structure. We developed a nanocomposite type hydrogel (NC gel) which solves all these problems simultaneously. The NC gels, were prepared by in situ free radical polymerization using exfoliated clay, instead of an organic crosslinker, and consisted of a unique organic (polymer)/inorganic (clay) network structure. NC gels exhibited high transparency, large and rapid swelling/de-swelling, and superb mechanical properties. Also, NC gels exhibit a number of interesting new characteristics related to the properties of their gel-air and gel-water interfaces, coil-to-globule transition, optical anisotropy, self-healing, harvesting cells, and control of morphology.1, 2 Herein, I present recent developments on the functions of NC gels; e.g., the synthesis of new type of NC gel with ternary PNIPA-clay-silica structure which exhibits anomalous increases in mechanical properties, 3 the generation of large retractive tensile forces in brine comparable to human muscle due to the salt-induced coil-to-globule transition of PNIPA, 4 and instant strong adhesive behavior of NC gel toward hydrophilic porous materials.5

INVITED LECTURE

# Preparation of hyperbranched multimethacylate polymers containing benzene and urethane moieties and its application in dental composites with combined high performance properties

Xiaoze Jiang
Donghua University

Dental restorative composite resins have been widely used instead of toxic amalgam alloys due to their characteristics of esthetics and functions. However, the polymerization shrinkage of organic matrix of composite resin is the main problem of this filed, which leads to the marginal gaps and secondary caries. To address this problem, novel hyperbranched multimethacrylate polymers (HBP) containing benzene, urethane, and vinyl moieties were synthesized from hydroxyl-terminated flowable hyperbranched polyamine-ester and isocyanoethyl methacrylate with the aim of replacing Bis-GMA as one component of dental restorative materials. The structure of HBPs was confirmed by FT-IR and 1H NMR spectra. HBPs were incorporated into Bis-GMA/TEGMA resin system to replace Bis-GMA partly and totally. Double bond conversion, polymerization volumetric shrinkage, water sorption and solubility, flexural strength and modulus of HBP containing resin formulations were investigated with conventional Bis-GMA/TEGMA based resin as the control. Results showed that the polymerization shrinkage of resin could be reached below 2% without compromising of their physical-chemical and mechanical properties by the optimal addition of HBPs (20 wt.%) into Bis-GMA/TEGMA system, comparing with those of control Bis-GMA/TEGMA based resin. Therefore, this study exhibits high potentials to prepare dental composite resin with low shrinkage and high strength properties.

INVITED LECTURE

# Study on the free radicals induced by gamma rays in polyacrylonitrile nano-fibers using electron spin resonance spectroscopy

Weihua Liu, Mouhua Wang, Guozhong Wu, Zhongfeng Tang
Shanghai Institute of Applied Physics

Polyacrylonitrile (PAN) is known to be the most commonly used precursor for producing carbon nano-fibers. The technique for producing PAN-based carbon nano-fibers is the stabilization in air and then carbonization of electrospun PAN nano-fibers in Ar/N2. Stabilization plays an important role in the performance of carbon nanofibers, which is essential, time-consuming and intensely exothermic. Researchers have paid much attention to optimizing the stabilization process by post-spinning modification to lower fusion and fiber-to-fiber adhesion. Radiation technique was found to be effective in moderating the exothermic behavior and accelerating stabilization process while the mechanism is not clear. Radiation induce formation of free radicals leading to crosslinking and degradiation and they are considered to be related to the modification mechanism. In this study, PAN nano-fibers were produced via electrospun technique and irradiated to various doses (0-500kGy) in air and vacuum by gamma rays. Then the structure and properties of the free radicals in PAN nano-fibers were investigated using electron spin resonance(ESR) in detail. The results indicate that mainly two types of radicals, namely alkyl radicals and polyimine radicals, were formed and the radical concentration reached saturation of 1.3*1019spins/g at 200kGy in vacuum at room temperature based on the ESR measurement. However, the radical concentration induced by radiation in air was rather low and only a weak sinlget was observed in the ESR spectra. This attributed to the presence of oxygen which easily reacted with the free radicals and consumed the radicals formed during irradiation process. The radical stability and decay behaviors at room temperature and elevated temperatures were also investigated under different atmospheres. It was found that the free radicals were rather stable when stored in vacuum at room temperature and the half-life period was longer than 140d. But the radicals rapidly decayed via reaction with oxygen when exposed to air and the radical concentration decreased to half of the initiation concentration within 20 minutes. The decay rate of free radicals increased by elevating temperature invacuum. The alkyl radicals disappeared completely at 110°C, but about 30% of the polyimine radicals remained indicating that polyimine radicals are more stable compared to the alkyl radicals due to their lower mobility.

ORAL PRESENTATION

# 2-D onjugated polymer nanosheets for photocatalytic overall water splitting

Lei Wang
Department of Polymer Science and Engineering, University of Science and Technology of China

In recent years, semiconducting conjugated polymers have emerged as a novel class of photocatalyst for solar-driven water splitting. Compared to their inorganic counterparts, these synthetic conjugated polymers offer great versatility to develop highly efficient photocatalyts with tunable electronic structures for photocatalytic applications. So far, much efforts have been devoted to developing polymer photocatalysts for photocatalytic hydrogen evolution. However, conjugated polymers that are able to efficiently split pure water under visible light (>400 nm) still remain to be explored. We show that 1, 3-diyne-linked 2D conjugated polymer nanosheets obtained by oxidative coupling of terminal alkynes such as 1, 3, 5-tris-(4-ethynylphenyl)-benzene (TEPB) and 1, 3, 5-triethynylbenzene (TEB) are possessing suitable band structures for photocatalytic overall water splitting and can act as highly efficient photocatalysts for splitting pure water (pH~7) into stoichiometric amounts of H2 and O2 under visible light irradiation. Using in situ techniques, we could further elucidate the reaction pathways during the photocatalytic process, providing strong evidence that the water splitting reaction could occur on the surface of polymer photocatalysts. Meanwhile, inspired by natural photosynthesis, Z-scheme photocatalytic systems are very appealing for achieving efficient overall water splitting. We also show the construction of polymer-based van der Waals heterostructures as metal-free Z-scheme photocatalytic systems for overall water splitting using aza-CMP and C2N ultrathin nanosheets as O2-evolving and H2-evolving catalysts, respectively. We believe that our study could provide new insights in design and synthesis of semiconductors that are able to catalyze overall water splitting at neutral pH with sunlight as the only energy input.

**POSTER PRESENTATION**

# Fabrication of the nerve agent detecting device with high sensitivity and selectivity

JiaNan Weng, Bo Zhu
Shanghai university

The detection of nerve agent was important in the modern warfare. To meet the requirement in the battlefield, high detecting sensitivity and selectivity were necessary. Herein, we report a new type of nerve agent detecting device by spin-nanoassembly of nerve-agent targeted PEDOT-HFIP on commercial interdigitated electrodes. At the optimized space-to-volume ratio, we demonstrate a low-power, sensitive, and selective nerve agent sensing technology using this platform by detecting GB vapor with a LOD of 10 ppb and a signal strength of 400 times of interference from the same concentration water, offering significant advantages over existing similar technologies.

**POSTER PRESENTATION**

# Multiple superwettable nanofiber arrays by a controllable dewetting strategy

Zhongxue Tang, Cong Lv, Huan Liu
Beihang University

Superwettable solid surfaces, fabricating rough surfaces and tailoring the surface chemistry, have attracted extensive attention, owing to their excellent performances. We proposed a facile dewetting strategy for fabricating multiple superwettable copper hydroxide nanofibers arrays NFAs ($Cu(OH)_2$-NFAs) by controlling the distribution states of silicone oil on the NFAs. It is proposed that the co-effect of the capillary force and the evaporation process can drive the localization of silicone oil on the NFA. With increasing the concentration of silicone oil, the discrete distribution of silicone oil on the NFA was turned to become a continuous thin/thick film distribution, which simultaneously changed the surface energy and roughness. By controlling the etching process, nanostructures and micro-/nano-structures were separately textured on the copper plates, resulting in the accurate manipulation of the surface adhesion depending on the transition of the superhydrophobicity state of the NFAs from Wenzel state to Cassis state. Herein, the controllable properties of the $Cu(OH)_2$-NFAs was realized, such as superhydrophilicity (SHPL), superhydrophobicity (SHPB), superslippery property (SPL), the adjustable adhesion, etc. It is worth noting that silicone oil can migrate to the top of the nanofibers during the dewetting process, leading to a SHPB NFA after dewetted from a very small amount of silicone oil. We envision that these results will provide new inspiration on the facile fabrication of functional surfaces with multiple superwettabilities.

POSTER PRESENTATION

# Scale-up preparation and application of microporous organic polymers

Lei Li, Tingting Zhu, Qi Yu, Hui Gao
Xiamen University

Microporous organic polymers (MOPs) with high surface area, low framework density and excellent thermochemical stability, have potential applications in gas adsorption, separation, energy storage, heterogeneous catalysis and sensor. However, expensive monomer and catalyst, harsh reaction condition, as well as complex preparation process limit their practical applications. Therefore, low-cost, scale-up and sustainable preparation of high performance MOPs still remains great challenges. In this report, low-cost building blocks (asphalt, divinylbenzene, pentaerythritol tetraacrylate and bismaleimide) were employed to construct MOPs through Friedel-Crafts reaction and free-radical polymerization. The resultant MOPs exhibit high surface area and excellent gas ($CO_2$, $H_2$ and VOCs) and dye adsorption capacities. The efficient and scale-up preparation of functional MOPs is meaningful for the future industrial practice.

POSTER PRESENTATION

# The structure and property study for PA6/'soluble' $TiO_2$ nanocomposite fiber materials

Hailong Yu[1], Feng Deng[2], Weidong Zhan[2]
1.college of material science and engineering of Donghua University
2.Zhejiang HuiBo New Materials Co., Ltd.

Nowadays, for improving the dispersion and compatibility between nanoparticles fillers and matrix such as PA6 resin, many researchers are attempting to study the relationship between structure and property of nanocomposite materials via interaction and interlayer methods. As one of main fillers, TiO2 is used widely in the industry of PA6 composites fibers materials. However, due to dispersion problem and worse compatibility the maximum TiO2 content among PA6 resin for fiber application is limited at low level. Via the study of interaction and interlayer for PA6/nano TiO2 composites can explore the relationship between structure and property and then improve dispersion and compatible performance for PA6/nano TiO2 composites. And it is also the direction to make new materials towards integration with structure, function and performance.

In order to study the influence of TiO2 surface modified with different functional group type of silane couple agent and how these 'soluble' fillers interact with polymer chain via different type of interlayers compared with commercial in-situ TiO2 composites, this work is using comparation methods to establish experimental samples. One side, after surface modified with silane coupling agent KH550 with NH3 amino group or KH570 with double bond group on nanoparticles TiO2, we prepared evenly mixed nanocomposites samples with different content of nano TiO2 based on PA6 matrix in melt blending process. Another side, the comparation samples were in-situ polymerization PA6 composites contained different percentage of TiO2 from market.

**KEYNOTE SPEECH**

# Bioinspired graphene-based nanocomposites

Qunfeng Cheng
Beijing University of Aeronautics and Astronautics

With its extraordinary properties as the strongest and stiffest material ever measured and the best-known electrical conductor, graphene could have promising applications in many fields, especially in the area of nanocomposites. However, processing graphene-based nanocomposites is very difficult. So far, graphene-based nanocomposites exhibit rather poor propertiess. Nacre, the gold standard for biomimicry, provides an excellent example and guidelines for assembling two-dimensional nanosheets into high performance nanocomposites. The inspiration from nacre overcomes the bottleneck of traditional approaches for constructing nanocomposites, such as poor dispersion, low loading, and weak interface interactions. Herein, we summarizes recent research on graphene-based artificial nacre nanocomposites, [1-5] and focuses on the design of interface interactions and synergistic effects for constructing high performance nanocomposites.

**KEYNOTE SPEECH**

# Structural regulation of graphene for fiber composite materials

Yongxiao Bai, Lijun Jin
Key Laboratory of Special Function Materials and Structure Design of Ministry of Education , MOE Key Laboratory for Magnetism and Magnetic Materials, Institute of Material Science and Engineering, Lanzhou University

Green, simple, low-cost method for preparing high-quality graphene by a simple and convenient method using natural honey as exfoliating media. And the plate diameter size and thickness or number of layers could be regulated via preparation technology and supplemental materials. Then, a series of polymer-graphene nanocomposite fibers were prepared by spinning. Natural polymer fiber is favored by academic and industrial circles because of its wide sources of raw material, pollution-free, recyclability, high moisture absorption. However, the low tensile strength of many fibers has also become a key bottleneck that hinders its further development. In order to expand the application range of natural polymer fiber, it is of great significance to prepare high-strength polymer fiber. The preparation of composite materials by nanocomposite is an effective method to improve the properties of the original materials. Fillers for composite materials are typically made of materials with outstanding properties. Graphene and Graphene Oxide (GO) are widely used in nanocomposites of polymer composite because of their excellent mechanical properties, antibacterial and ultraviolet radiation resistance. In this talk, the preparation of high-strengh composite polymer fiber was taken as the research object and GO was used as a nano-filler. The GO/polymer composite fiber was prepared by solution compounding and wet spinning method. Then the cross-linking modification method was adopted to further improving the performance of the fiber. The relationship between GO morphology, crosslinking conditions and performance of the composite fiber was studied. Trying to improve the overall performance of polymer fibers especially solving the basic problems of low strength of general fibers.

INVITED LECTURE

# Deformable structures for flexible smart materials

Zunfeng Liu
Nankai University

Superelastic conducting composites with improved properties and new functionalities are needed for applications ranging from electronic interconnects to sensors and electrically driven artificial muscles. In this talk, I will present our work using hierarchically buckled sheath-core structures based on carbon nanotube sheets and rubber fiber. Various applications such as superelastic conducting fibers, resistive and capacitive strain sensors, stretchable antennas and artificial muscles will be demonstrated. I will further show their applications in wearable healthcare devices such as stretchable electrodes for glucose detection, body temperature sensor, artificial muscle, and inflatable electrocauterization devices. In addition, by introducing twisting technology into the fiber, the stretching stroke of the artificial muscle fiber is greatly increased, and a moisture-sensitive smart fabric is constructed. The smart fabric is woven by the stretching muscle of silk to realize the stretching of the length of the smart fabric when the environmental humidity changes. Besides, the double mechanism of thermal expansion and infrared absorption is also used to achieve the overspeed movement of artificial muscle. The water-sensitive textile can realize the management function of water and heat by changing the shape and structure, which is expected to bring new opportunities to the intelligent textile field.

INVITED LECTURE

# Smart clothing materials

Chengyi Hou
Donghua University

In recent years, with the rapid development of novel technologies such as artificial intelligence, internet of things and cloud computing, research on smart clothing which is considered as the new medium that connects human body with technologies, has drawn wide attention from academic and industry. Our research group (pilab.dhu.edu.cn/afmg) has dedicated to the research of flexible functional materials for smart clothing for many years. The achievements include: a) flexible chromic materials and devices, b) flexible sensing devices, c) flexible energy devices such as solar cells, thermoelectric, triboelectric and supercapacitors, d) flexible circuits, e) soft actuators. Some recent published works will be introduced in this talk.

INVITED LECTURE

# Hybrid energy fabric based on fiber electrodes for wearable electronics

Nannan Zhang, Changyuan Tao, Xing Fan
Chongqing University

The demand for wearable power sources has been growing rapidly in the modern era of portable electronics. It is essential to make these power sources more comfortable and fit for the human body, especially for some medical applications in which case we need to wear it day and night. Energy harvesting and storage devices in the form of a common-looking fabric is thus of a great interest in recent years. Herein, we have successfully woven out a type of micro-cable power textile for simultaneously harvesting energy from ambient sunshine and mechanical movement. Solar cells fabricated from lightweight polymer fibers into micro cables are then woven via a scalable shuttle-flying process with fiber-based triboelectric nanogenerators to create a hybrid energy harvesting fabric. The hybrid power fabric is just 320 μm thick, fabricated with a size of 4 cm by 5 cm, was demonstrated to charge a 2 mF commercial capacitor up to 2 V in 1 min under ambient sunlight in the presence of mechanical excitation, such as human motion and wind blowing. Based on the similar strategy, a novel solid solar energy harvesting and storage hybrid textile combining both photovoltaic cells and super-capacitors were also designed and realized, in order to provide power at dark condition within a single textile. With such features as low-cost, thinness, light-weight, being flexible and wearable, as well as colorful appearance, this work represents a significant step towards the development of large scale energy harvesting and storage for wearable electronics and self-powered home technology.

**ORAL PRESENTATION**

# Electrospun core-shell nanofiber with electric triggered shape memory behavior

Yuliang Xia, Zhang Fenghua, Liu Yanju, Leng Jinsong
Harbin Institute of Technology

Shape memory polymers, as a class of smart materials, can transform between the temporary shape and permanent shape. This temporary shape can be maintained below the glass transition temperature steadily or switch to the permanent shape with external stimulation. Using electric current to trigger the shape memory effect is an ideal approach cause it allows the remote shape transformation control of the polymer. Conductive fillers such as carbon nanotube, carbon black, and carbon fiber have good conductivity, but these materials tend to aggregate during the polymer crosslinking process, which will reduce the polymer composite conductivity. Herein we combined the shape memory poly-lactic acid (PLA) with conductive poly pyrrole (PPy) to create a conductive shape memory composite membrane. Electrospinning technique was employed to form shape memory PLA nanofiber membrane. The PPy shell was coated on the PLA nanofiber surface by chemical vapor deposition to form a conductive network. The PLA-PPy core-shell structure was observed under the scanning electron microscope. Different factors that affect the chemical vapor deposition process were studied. Furthermore, we characterized the electrically triggered shape memory behavior of the nanofiber membrane. The nanofiber membrane can revert from a bent temporary shape to the original shape in 2s. Surface temperature distribution during the electrified shows a uniform conductive fiber distribution. The electric triggered shape memory property of the PLA-PPy membrane allows it to be used in smart material applications.

**ORAL PRESENTATION**

# A highly efficient self-healing, super tough elastomer for stretchable devices

Luzhi Zhang, Zenghe Liu, Zhengwei You
Donghua University

It is highly desirable, although very challenging, to develop self-healable materials exhibiting both high efficiency in self-healing and excellent mechanical properties at ambient conditions. Herein, a novel Cu(II)–dimethylglyoxime–urethane-complex-based polyurethane elastomer (Cu–DOU– CPU) with synergetic triple dynamic bonds is developed. Cu–DOU–CPU demonstrates the highest reported mechanical performance for self-healing elastomers at room temperature, with a tensile strength and toughness up to 14.8 MPa and 87.0 MJ m−3, respectively. Meanwhile, the Cu–DOU–CPU spontaneously self-heals at room temperature with an instant recovered tensile strength of 1.84 MPa and a continuously increased strength up to 13.8 MPa, surpassing the original strength of all other counterparts. Density functional theory calculations reveal that the coordination of Cu(II) plays a critical role in accelerating the reversible dissociation of dimethylglyoxime–urethane, which is important to the excellent performance of the self-healing elastomer. Application of this technology is demonstrated by a self-healable and stretchable circuit constructed from Cu–DOU–CPU.

**KEYNOTE SPEECH**

# Silk flexible electronics: meso-construction of biocompatible silk-hybrid materials to acquire extraordinary performance and smart sensing

Xiangyang Liu
Department of Physics, FOS, National University of Singapore, Singapore College of Physical Science and Technology, Xiamen University

The design and fabrication of bio-mimicking materials has been found to have broad applications in artificial skins, health monitoring, soft robotics, etc. Herein, we report a super-sensitive strain sensor based on highly elastic and self-recovery wool keratin gels. The keratin gels can be obtained by chemically crosslinking the cysteine residues of extracted keratins with 4-arm poly(ethylene glycol) vinyl sulfone. Since a large number of inter-connected α-helices by S-S bonds inside the wool keratin chemical gels, they behave as three-dimensional (3D) molecular-scale springs, which display superior recovery capability in distortion. This gives rise to the ability to withstand more than 10, 000 stress-strain cycles with an extreme low dissipation energy (very narrow hysteresis loop). Intriguingly, the elastic properties of 3D molecular springs highly depend on the chemical crosslinking density in the network structures. As a keratin gel mediated strain sensor, the dynamic range of this sensor can be tuned easily by chemically adjusting the crosslinking degree of the gel network, so that the device can be tuned to detect a subtle strain with high sensitivity. Such a senor has also been applied to discern voice, detect and track wrist pulse and the movement of bodies

As an excellent flexible biomaterial, Bombyx mori silk fibroin (SF) materials offer exquisite mechanical, optical, and electrical properties which are advantageous toward the development of next-generation biocompatible electronic devices. In this concern, the functionalization of SF materials can be implemented by re-engineering the the hierarchical structure of SF materials, which endows the materials with the particular properties applied to the fabrication of flexible electronics and smart sensors. This presentation will outline the strategy from meso re-construction of the hierarchical structure of Bombyx mori silk materials to acquire some particular functionalities based on Intra/inter molecular nucleation. The principles and examples of designing and integration of high-performance bio-integrated devices will be outlined for future applications in consumer, biomedical diagnosis, and human–machine interfaces.

INVITED LECTURE

# 碳纳米管纤维的连续制备及高性能化技术

张永毅
中国科学院苏州纳米技术与纳米仿生研究所南昌研究院

碳纳米管具有长径比超高、力学性能超强、综合性能优异等特点，被认为是制备超强纤维的理想基元。然而，从碳纳米管出发构建碳纳米管纤维是一个从一维到三维、从微观到宏观的复　杂过程，其间性能跨纬度、跨尺度传递问题突出。浮动催化法是一种工艺简单、易于实现低成本连续化组装的碳纳米管纤维制备技术。本报告将围绕碳纳米管纤维的连续制备及高性能化这一主题，讲述所在团队在近年在碳纳米管纤维的浮动催化法连续制备、碳纳米管的微观结构调控及纤维的连续高性能化等方面的研究进展。

INVITED LECTURE

# Biomimetic thermo-responsive fibrous hydrogels: highly mechanical strength and ultra-fast response underwater

Qingsong Zhang, Qifeng Mu, Wen Yu, Li Chen, Xiaoyun Liu
1.Tiangong University
2.Donghua University

In striking contrast to most synthetic hydrogels, biological hydrogels are composed of weakly crosslinked fibrous networks. The unique fibrous architecture gives rise to special properties, such as low concentration, high porosity and anisotropic hydrogels with a high rate of substance exchange and a recoverable energy hysteresis. Here, we demonstrate a facile method for the photo welding of fibrous mats of electrospun nanofibers by introducing a photo-crosslinker 4-acryloyxybenzophenone (ABP) into the micro-/nanofibers. By leveraging the strong capillary effect of the porous mats, the welded fibrous networks can transform into hydrogels with biomimetic networks inside upon immersed in water. This method allows us to lock the semiflexible fibrous networks present at the welding conditions, and these junctions ensure the high stability of fibrous networks in water over one week. Through facilely tuning the conformation of molecular chains by temperature, the synergistic effect of hydrophobic associations and hydrogen bonding, serving as reversible sacrificial bonds for energy dissipation, impart the fibrous hydrogels with fracture strength (~0.38 MPa) and toughness (~1560 J/m2). Applying the fibrous hydrogels for protein sorption, ultrafast sorption/desorption can be achieved due to the thermo-responsive substrates with high specific surface area and the multiscale size-exclusion effect. This study opens up a very promising avenue for obtaining ideal biomimetic materials as underwater smart biological films or artificial skins.

INVITED LECTURE

# Intelligent bidirectional thermal regulation of phase change material incorporated in thermal protective clothing

Yun Su[1,2,3], Wen Zhu[1], Xianghui Zhang[1,2], Jun Li[1,2]
1.Donghua university
2.Key Laboratory of Clothing Design and Technology, Donghua University, Ministry of Education
3.Shanghai center for high performance fibers and composites, Shanghai
4.Center for Civil Aviation Composites, Donghua University, Shanghai

Coating of fabric by phase change materials (PCMs) changed thermal properties of the fabric. The PCM coated fabric with intelligent bidirectional thermal regulation was designed and prepared. The basic and thermal physical properties of the PCM coated fabric were measured. The PCM layer was incorporated into thermal protective clothing, and the thermal protective performance was evaluated under hot contact exposure. High thermal capacity of PCM increased the potential of thermal protective clothing for heat accumulation during heat exposure, but also modified heat release from the thermal protective clothing during cooling. The incorporation of the PCM coated fabric increased greatly the thermal protective performance during the heat exposure, which was influenced by melting temperature, PCM content and enthalphy. The thermal hazardous effect caused by the thermal protective clothing was slightly increased, but presenting no significant correlation with the skin absorbed thermal energy during the cooling. Therefore, the PCM showed great potential applications for developing intelligent thermal protective clothing. The conclusions obtained from this study contributed to development of PCM with high thermal capacity and low heat release suitable for the thermal protective clothing.

**ORAL PRESENTATION**

# Photoreversible colorswitching of rewritable smart fabrics driven by UV/NIR light

Daniel Kinyanjui Macharia, Zhigang Chen
State Key Laboratory for Modification of Chemical Fibers and Polymer Materials, International Joint Laboratory for Advanced Fiber and Low-Dimension Materials, College of Materials Science and Engineering, Donghua University, Shanghai 201620, P. R. China

Remote, rapid, and ink-free printing/erasure on fabrics has great potential to revolutionize specialized clothing in numerous applications including fashion/aesthetic and security fields, but the construction of such smart fabrics has not been realized due to underlying obstacles in obtaining suitable photoreversible color-switching systems (PCSS). To address this problem, we have prepared $TiO_{2-x}$ nanorods as photocatalytic and photothermal component. With redox dyes as reversible color indicators and hydroxyethyl cellulose (HEC) as polymer matrix, $TiO_{2-x}$/dye/HEC-based PCSS is coated on poly(dimethylsiloxane)-treated cotton fabric. Under 365 nm light irradiation, discoloration occurs in 180 s, resulting from the efficient photocatalytic reduction of the dye. On the contrary, when the colorless fabric is irradiated by 808 nm light, recoloration occurs in a very short time (~ 100 s), far lower than the traditional heating mode (30–8 min at 90–150 °C). This rapid recoloration should be attributed to the localized high temperature (164.3–184.5 °C) induced by photothermal effect of $TiO_{2-x}$. Particularly, when $TiO_{2-x}$/dye/HEC-based PCSS is extended to coat commercial clothes (such as T-shirts), red/green/blue figures/letters can be rapidly and remotely printed by UV-light pen and then erased by near-infrared light, with high cycle stability. Therefore, such rewritable smart fabric represents an attractive alternative to regular clothes in meeting the increasing aesthetic or camouflage needs.

**KEYNOTE SPEECH**

# 无卤阻燃锦纶及其混纺复合纺织材料的研究

付常俊[1]，付欧丽[1,2]，古哈尔娅 伊孜丁[2]，彭志汉[2,2]，袁丽菊[1]，王朝生[1]，郁崇文[2]
1.上海安凸阻燃纤维有限公司
2.东华大学

当前锦纶阻燃是通过添加含卤或无卤阻燃胶表面涂覆等后整理方式实现，虽然缓解了市场对阻燃锦纶的需求，但对环保和织品重量等存在巨大设计缺陷，尤其是自 ROHS2.0，REACH SVHC 清单，OEKO-TEX® STANDARD 100 等环保认证和法律法规的实施，新型环保、安全的阻燃和开发应用，阻燃纤维的无卤化、抑烟及减毒已经成为阻燃研究的前沿性课题。

锦纶本身纺丝粘度高、无机纳米助剂分散困难的特性，尤其的在有机/无机混杂的助剂体系尤为困难。锦纶无卤阻燃体系的助剂，耐温性差、易分解，Td 接近加工温度上限，导致加工窗口窄，工业化纺丝稳定性难度极高。 相对于“原位聚合阻燃锦纶切片”方法、 “多元阻燃单体共聚锦纶”方法，本研究所采取的“共混熔融阻燃锦纶母粒纺丝”无需改变聚合阶段的工程装备 。

1）解决了锦纶阻燃体系中“有机+无机”混合助剂间的超分散关键技术，制备了高可纺性阻燃母粒 ，用共混熔融纺丝的方式，实现了无卤阻燃锦纶单丝、长丝（FR-PA6，FR-PA66）和 BCF 纱等核心产品的规模化生产，克服了纺丝过程中锦纶结晶和取向特性的无序波动的困难，实现连续稳定的纺丝生产长丝，产品的机械性能、阻燃性能、燃烧烟毒性烟密度等特征实现重大突破和成功应用；该类产品 CESALON® A　阻燃锦纶 BCF 纱替代传统的羊毛绒头纱生产航空地毯，具有重量轻（每架飞机节省 100-120KG 地毯和重量），耐用（比传统羊毛航空地毯寿命高一倍以上），同时具有极低的烟密度和烟毒性，远远超过 FAR25.853 的航空内饰烟毒性和烟密度要求。是制备新一代民用航空地毯和座椅套的核心材料，在军警消防用织带、便携具、降落伞面料等获得应用。
（2）运用“灯芯效应阻断”原理制备的阻燃锦纶交联型母粒，创新性突破了“热固+热塑”多元纤维复合的“灯芯效应”的必然障碍，阻燃锦纶 CESALON® X 系列短纤，实现了和聚酰亚胺、阻燃粘胶、阻燃棉、芳纶等热固性纤维的混合纱线生产，实现无融滴、无续燃复合纱线而不产生阻燃冲突性 ，综合应用耐用性、舒适性和其它性能互补，成为阻燃工装各种面料、作战服面料、个体防护面料和防护悬挂织带等克选用的新一代核心材料。

KEYNOTE SPEECH

# The evolution of strong, fast, powerful, durable, and cheap polymer artificial muscles from carbon nanotube muscles

Shaoli Fang
the University of Texas at Dallas

Twist-spun carbon nanotube (CNT) yarns derived from CVD-grown spinnable multiwalled nanotube forest are of great interest for such diverse applications as artificial muscles, supercapacitors, batteries, intelligent textiles and structural composites. These twist-spun, torque stabilized nanotube fibers can be produced with strength over 700 MPa. Remarkable performance has been obtained for tensile and torsional CNT hybrid yarn muscles, whose actuation is driven by the volume change of a guest within a twisted or coiled carbon nanotube yarn. During thermally-powered contraction, coiled hybrid muscles can deliver 29 times the work as the same weight human muscle. However, CNT yarn muscles are relatively expensive and not widely available. We here report artificial muscles using fishing line, sewing thread, and other natural fibers such as silk fibers, which are cheap and widely available. Our coiled polymer muscles exhibit outstanding muscle performances, providing higher gravimetric torque than for large electric motors and rotates much heavier rotors at over 100, 000 rpm and tensile & torsional work capacities of 2.5 kJ/kg and 2.1 kJ/kg, compared with 0.93 kJ/kg for tensile actuation of NiTi shape memory wires. A 0.6 g muscle, made by coiling an 800-μm diameter polyethylene fishing line, lifts a 7.2-kilogram weight by 2.2 cm twice per second when powered by a 70°C temperature change (providing 7.1 horsepower per kg). A 25°C temperature change provides tensile stoke of 30%, yielding a stroke coefficient of 1.2%/°C. Using homochiral and heterochiral coiled configuration, muscles can either contract or elongate when heated. Muscles using silk fibers, which actuate in response to high moisture, are also discussed.

**INVITED LECTURE**

# Silk-based electronic fibers and textiles

Yingying Zhang
Tsinghua University

The development of flexible electronics and equipment attracts significant interests in recent years. It is of great importance to explore low cost and scalable preparation approaches for high performance flexible and wearable electronics. Silkworm silk, with five thousand years' usage history, is a popular natural material for clothes or wearing accessories. In this talk, I will present our work on exploring the application of silk fiber/fabrics in flexible electronics. We demonstrated carbonized silk fabric could be worked as strain sensors with both of high sensitivity and high tolerable strain for monitoring both of vigorous human motions, subtle human motions and even sound, showing their superior performance and tremendous potential applications in wearable electronics and intelligent robots. We also demonstrated the fabrication of high performance silk E-skin, silk/graphene E-tattoos, and silk based 3D printed silk-based smart fibers/textiles. These work may help to extend the application of silk materials from traditional textile industry to next-generation wearable electronics and smart textiles.

INVITED LECTURE

# A wearable, thread/paper-based microfluidic device coupled with smartphone for sweat analysis

Zhisong Lu, Gang Xiao, Yan Qiao, Ling Yu
Southwest University

Wearable devices have received tremendous interests in human sweat analysis in the past few years. However, the widely used polymeric substrates and the layer-by-layer stacking structures greatly influence the cost-efficiency, conformability and breathability of the devices, further hindering their practical applications. Herein, we reported a facile and low-cost strategy for the fabrication of a skin-friendly thread/paper-based wearable system consisting of a sweat reservoir and a multi-sensing component for simultaneous in situ analysis of sweat pH and lactate. In the system, hydrophilic silk thread serves as the micro-channel to guide the liquid flow. Filter papers were functionalized to prepare colorimetric sensors for lactate and pH. The smartphone-based quantitative analysis shows that the sensors are sensitive and reliable. Although pH may interfere the lactate detection, the pH detected simultaneously could be employed to correct the measured data for the achievement of a precise lactate level. After being integrated with a hydrophobic arm guard, the system was successfully used for the on-body measurement of pH and lactate in the sweats secreted from the volunteers. This low-cost, easy-to-fabricate, light-weight and flexible thread/paper-based microfluidic sensing device may hold great potentials as a wearable system in human sweat analysis and point-of-care diagnostics.

INVITED LECTURE

# Highly efficient self-healable and dual responsive hydrogel-based deformable triboelectric nanogenerators for wearable electronics

Qingbao Guan, Zhengwei You
Donghua University

Self-healable soft conductors, which can withstand certain degrees of deformation and can recover from damage spontaneously, are essential for wearable applications. In this work, a soft hydrogel based self-healing triboelectric nanogenerator (HS-TENG), which is highly deformable, both mechanically and electrically self-healable, has been successfully fabricated from poly(vinyl alcohol)/agarose hydrogel. The incorporation of photothermic-active polydopamine particle and multiwalled carbon nanotube (MWCNT) allows the HS-TENG to be physically self-healed in ~1 min upon the exposure of near-infrared (NIR) light. At the same time, the chemical self-healing of HS-TENG can be triggered by water spraying at 25 °C when introducing water-active dynamic borate bond into hydrogel. The applicability of HS-TENG as a soft energy device to harvest human motion energies has been demonstrated. This work provides a feasible technology to design TENG with dual self-healing mode, which extends the reliability of TENG for wearable electronics and potentially solve the energy issues of soft electronics.

INVITED LECTURE

# Core–shell-yarn-based triboelectric nanogenerator textiles as power cloths

Junyi Zhai
Beijing Institute of Nanoenergy and Nanosystems, CAS

Although textile-based triboelectric nanogenerators (TENGs) are highly promising because they scavenge energy from their working environment to sustainably power wearable/mobile electronics, the challenge of simultaneously possessing the qualities of cloth remains. In this work, we propose a strategy for TENG textiles as power cloths in which core–shell yarns with core conductive fibers as the electrode and artificial polymer fibers or natural fibrous materials tightly twined around core conductive fibers are applied as the building blocks. The resulting TENG textiles are comfortable, flexible, and fashionable, and their production processes are compatible with industrial, large-scale textile manufacturing. More importantly, the comfortable TENG textiles demonstrate excellent washability and tailorability and can be fully applied in further garment processing. TENG textiles worn under the arm or foot have also been demonstrated to scavenge various types of energy from human motion, such as patting, walking, and running. All of these merits of proposed TENG textiles for clothing uses suggest their great potentials for viable applications in wearable electronics or smart textiles in the near future.

**INVITED LECTURE**

# Soft actuators and soft robots

Jian Zhu
National University of Singapore University

Robots have been proposed to relieve human beings from dangerous environments or tedious manufacturing processes, or mimic human beings or animals in appearance, behaviour, and/or cognition. Most of traditional robots use hard materials. Inspired by natural creatures, researchers recently become more and more interested in soft robots, which are made of soft materials, say polymers.

To exploit animal-like locomotion and behaviour, it is essential to develop muscle-like actuators with properties close to biological systems. We employ dielectric elastomers - one class of soft active materials - as artificial muscles in soft robots. A dielectric elastomer actuator can deform in response to voltage, and can exhibit unique attributes, including large voltage-induced deformation, fast response, low weight, quiet operation, etc.

In this seminar I will discuss several soft robots which are developed based on dielectric elastomer actuators, such as a jellyfish robot, artificial muscles for facial expressions, an inchworm-like robot, etc.

INVITED LECTURE

# Shear printed semiconductor electronics for smart fabrics

Gang Wang
Donghua University

The printed soft semiconductor electronics are expected to be one of the key parts for next-generation smart fabrics, attributing to the intrinsically wearable characters, solution printing process and unique logical response properties. Solution-based film processing techniques capable of increasing charge carrier mobility, electrical uniformity, and power conversion efficiency by molecular chain alignment, aggregation control and domain purity enhancement typically rely on mechanical stretching, high-temperature rubbing, or shear-printing. Among these methodologies, shear-printing has been of primary interest due to its applicability to rational microstructure design, soft/smart manufacturing process, and enhancement of charge transport/power conversion characteristics. In this presentation, the shear printing techniques are developed to realize the large-scale fabrication of semiconductor electronics for the applications in smart fabrics. The topics will include: 1. Microfluidic flow for the continuous and large-scale fabrication of semiconductor nanofibers; 2. New shear printing concept, including the nature-hair brush printing and mixed-flow design, for high-performance thin-film transistors, diodes and solar cells; 3. The integration of semiconductor electronics for smart fabrics and internet of things.

ORAL PRESENTATION

# Ionogel-based, highly stretchable, transparent, durable triboelectric nanogenerators for energy harvesting and motion sensing over a wide temperature range

Lijie Sun, Zhengwei You
Donghua University

Hydrogel-based triboelectric nanogenerators (H-TENGs) have shown great promise in wearable electronics as soft, stretchable and sustainable power sources. However, H-TENGs can only be used in a narrow temperature range for a short duration due to freezing and evaporation of water. Here, an ionogel-based triboelectric nanogenerator (I-TENG) is designed to significantly broaden the application temperature range and duration while retaining all the superior properties of H-TENGs. The ionogel network constructed by dipole-dipole and ion-dipole interactions exhibits high stretchability (~800%) and ionic conductivity (1.1 mS cm-1). The corresponding I-TENG retains high stretchability (>400%), transparency (>90%), and anti-fatigue resistance (resisting 1000 cycles of 100% stretching) with stable electronic performance for 1 month. The I-TENG shows an instantaneous peak power density of 1.3 W m-2 and efficiently harvests biomechanical energy to drive an electronic watch. Additionally, the I-TENG serves as a self-powered human motion sensor to inspect the bending angle of an elbow. More importantly, the I-TENG retains high stretchability and electrical performance over a wide temperature range from -20 to 100 °C. This work provides a new strategy to design and tailor TENGs that will be very useful for diverse applications, including wearable electronics, electronic skin, and artificial intelligence.

KEYNOTE SPEECH

# Smart wearables: nanotechnology & nanofibers

Seeram Ramakrishna
National University of Singapore

For thousands of years commodity fibers (natural polymers and synthetic polymers) are used by humans for garments, apparels, so on. Since 1950s high performance fibres (synthetic) with superior physical, mechanical, chemical and electrical properties enabled engineering applications. Future applications need multi-functional fibers with higher specific mechanical, chemical, biological, electronic, and information properties. Moreover, the need for Earth environment friendly textiles and processes on the backdrop of UN sustainable development goals, SDGs or circular economy vision of nations and companies is catalyzing fiber innovations. Objective of this presentation is to relate advances in electrospinning technology to the multi-functional fibers development for smart wearable applications and innovations. This lecture provides an overview, and discusses emerging opportunities in terms of high performance wearables, healthcare textiles, light weight textiles, self-cleaning textiles, anti-microbial textiles, smart wearables, electronic textiles, sensitive robot surfaces, edible textiles, sustainable textiles, regenerative textiles, and outer space wearables. Impact these innovations on the circularity of textiles and future living and well-being of humans will also be discussed. Emerging opportunities and challenges will also be deliberated.

**POSTER PRESENTATION**

# 聚丙烯酸修饰的氧化锌玻璃毛细管用于增强肿瘤标志物免疫荧光检测

武志华[1,2]，李耀刚[1]，段友容[2]
1. 东华大学
2. 上海市肿瘤研究所

氧化锌（ZnO）由于其突出的荧光增强效果而经常用作蛋白质微阵列基底。然而，功能性基底与微流体技术的整合以检测癌症标志物仍然需要被优化和促进，例如微/纳米结构的优化和用于免疫荧光测定的亲水性修饰策略等。在此基础上，我们采用微流体化学方法在玻璃毛细管内壁上构建 ZnO 纳米棒阵列，采用静电逐层自组装技术，用亲水性聚电解质 - 聚丙烯酸（PAA）对纳米棒阵列表面进行改性。并研究了流速和反应液浓度对 ZnO 纳米棒阵列形貌的影响。基于 ZnO 纳米棒阵列的玻璃毛细管，在 25 μL•min$^{-1}$流速下以 50 mM Zn$^{2+}$溶液制备 4 分钟，显示出最强的荧光增强性能。此外，PAA 的引入抑制了非特异性蛋白质的干扰，并有效地提高了抗体负载能力。基于此，在癌胚抗原（CEA）的检测中，检测限达到 100 fg•mL$^{-1}$，表明 ZnO @ PAA 纳米棒阵列为基础的微流体装置对蛋白质标志物具有显着的荧光增强及检测性能，具有应用于即时诊断和高通量癌症生物标志物检测的潜力。

POSTER PRESENTATION

# Prepolymerization-assisted fabrication of an ultrathin immobilized layer to realize a semi-embedded wrinkled AgNW network for a smart electrothermal chromatic display and actuator

Hongwei Fan
Donghua University

An ultrathin PDMS-immobilized layer was prepared to construct a semi-embedded wrinkled AgNW network and overcome the limitations through spin-coating the pre-polymerized PDMS solution. As a result, the as-prepared transparent stretchable AgNW/PDMS composite films with different AgNW loadings demonstrate great transmittance, conductivity, tensile stability under 40% and 60% strains, and adhesion of AgNWs on the PDMS substrate. With a further increase of AgNW loadings, reflective stretchable conductive films were obtained and they showed a much lower sheet resistance (~0.2 Ω sq-1) and good tensile stability under 70% strain. Based on these stretchable conductive films, stretchable electrothermal chromatic films and electrothermal actuators were fabricated to demonstrate their multifunctional applications.

**POSTER PRESENTATION**

# ZnS-CdS-TaON nanocomposites with enhanced stability and photocatalytic hydrogen evolution activity

Lin An
Donghua University

In recent years, tantalum oxynitride (TaON) semiconductor as one of the most efficient photocatalysts has been studied intensively owing to the appropriate potentials for overall solar water splitting. In this work, a stable ZnS/CdS/γ-TaON composite photocatalyst with wide visible light response is prepared by calcining γ-TaON at 800 °C in a wet $NH_3$ flowing and ion-exchange route in the presence of Cd, Zn and S precursors. The results show that the highest $H_2$ evolution rate of ZnS/CdS/γ-TaON composite is about 14 times higher than that of CdS/γ-TaON in the absence of any noble metal co-catalyst and 47 times higher than that of Pt loaded γ-TaON sample. The enhanced photocatalytic activity is attributed to the higher separation efficiency of electrons and holes with the heterogeneous junction between ZnS, CdS and γ-TaON. Besides, the presence of γ-TaON and ZnS also prevent CdS from the photocorrosion. The results provide a new insight into visible-light driven water splitting $H_2$ production of TaON-based nanocomposite photocatalysts.

POSTER PRESENTATION

# Nanoshell-wrapped carbon nanotube-SnS2 (MoS2) composites for flexible, stretchable and water-proof asymmetric supercapacitors

Qiufan Wang
South Central University for Nationalities

Manufacture of high-performance wearable supercapacitors requires a new class of flexible electrodes with high porosity, high conductivity, high mechanical stability and good water-proof ability. We successfully prepared a range of nanoshell-wrapped carbon nanotube-$SnS_2$ composite electrodes with excellent pseudo-capacitive characteristics. One of the as-prepared electrodes, $SnS_2$/CNT-8, exhibits an areal capacity of 850 mF $cm^{-2}$ at 10 mV $s^{-1}$. All-solid-state flexible asymmetric supercapacitor (ASC) based on $SnS_2$/CNT and $MoS_2$/PEDOT/CNT electrodes exhibited high performance with an extended potential window of 1.7 V, areal capacitance of 103.76 mF $cm^{-2}$ at 1.5 mA $cm^{-2}$, and outstanding stability with no capacitance degradation under a wide range of bending conditions. The ASC was sealed by polyimide films to achieve waterproof, which showed high electrochemical stability in hot water and under high speed centrifugation conditions, indicating good water-washability and potential wearability. The as-prepared ASC was also encapsulated in elastic films to provide 225% stretchabilty. The ASC devices packaged in all these ways exhibit high capacitance retention (> 90%) under various bending and dynamic conditions. These results indicate that the ASC can meet a wide range of energy output requirements

**POSTER PRESENTATION**

# A simple and high efficient approach to improve mechanical strength of carbon nanotube fibers via UV-initiated click reaction

Daohong Zhang, Xuan Ran, Shuangyuan Wang, Jiang Zhao, Qiufan Wang, Juan Cheng
South-Central University for Nationalities

Carbon nanotube (CNT) fibers have high potential for application in flexible electronics. However, it is still a challenge to produce CNT fibers with highly stable mechanical performance. Here, we use a high-efficient and facile thiol-ene click reaction to functionalize the CNT fibers so as to form a crosslinked structure between individual carbon nanutubes without damaging the structure of the CNTs. Several mercaptans with different functionalities were used to functionalize the CNT fibers via the thiol-ene click reaction between the double bond of pure CNT fibers (PCNTs) and the thiol groups. The effects of the type and content of mercaptans and reaction conditions on the performance of functional CNTs fibers have been studied in detail. The tensile strength of the functionalized CNT fibers increased with the increase of thiol functionality. The functionalized and twisted CNT fiber demonstrated 191.7% higher tensile strength than the PCNT. The enhancement mechanism is explained. This simple method about improving mechanical strength of CNTs fibers will assist their wide application.

POSTER PRESENTATION

# Bioinspired strong graphene-based nanocomposites for flexible energy storage

Tianzhu Zhou[1, 2], Chao Wu[2], Qunfeng Cheng*[1]
1.Key Laboratory of Bio-inspired Smart Interfacial Science and Technology of Ministry of Education, School of Chemistry, Beijing Advanced Innovation Center for Biomedical Engineering, BUAA-UOW Joint Research Centre, Beihang University, 37 Xueyuan Rd, Haidian District , China
2.School of Transportation Science and Engineering, Beihang University, 37 Xueyuan Rd, Haidian District , China

Flexible energy storage with outstanding property are needed to meet the increasing demand for wearable and portable electronics. There is a challenge to design exceptionally flexible device with strong graphene-based nanocomposites. Natural nacre shows remarkable fracture toughness because of its alternating inorganic/organic layered structure and abundant interfacial interactions, providing an inspiration for designing flexible energy storage. Here, we demonstrated nacre-inspired flexible supercapacitors through synergistic interfacial interactions of π-π stacking interactions, hydrogen bonding, and electrostatic interaction between halloysite (HA)-polyaniline (PANI) nanocomposites and graphene oxide (GO) nanosheets (HPA-rGO). The resultant graphene-based nanocomposites show strong tensile strength (351.9 MPa), high electrical conductivity, and long cycle life. Moreover, the assembled all-solid-state supercapacitors based on nanocomposites can not only presnet extraordinary flexibility with no decay of capacitance, but also deliver remarkable mass energy density up to 16.3 Wh kg<sup>-1</sup>. This nacre-inspired strategy for designing flexible nanocomposites provides an avenue for the next-generation power source in the fields of smart wearable electronics.

POSTER PRESENTATION

# Fabrication and characterization of bio-memristor based on silk fibroin

Yi Zhang[1, 2], Suna Fan[1, 2], Yaopeng Zhang[1, 2], Huili Shao[1, 2]
1.State Key Laboratory for Modification of Chemical Fibers and Polymer Materials
2.College of Materials Science and Engineering, Donghua University

As the fourth basic circuit component except resistors, inductors and capacitors, memristor is regarded as the next-generation non-volatile memory technology, which has the advantages of fast computing speed, high storage density, low power consumption and easy integration. It is expected to be applied to neural bionic construction, but its biocompatibility needs to be further improved. As a kind of biodegradable biomaterial with good biocompatibility, mechanical properties and optical properties, silk fibroin has broad application prospects in wearable, biosensing, storage and biomedical applications. However, the silk fibroin based memristor still has problems such as unstable memristive performance, and short retention time. In this paper, silk fibroin was used as the basic building material and composite metal ions. The flexible biological memristor with good biocompatibility and memristive performance was prepared to explore the memristive mechanism of silk fibroin materials and pave a new route for the development of new bio-memristors.

POSTER PRESENTATION

# A highly ionic conductive poly(methyl methacrylate) composite electrolyte with garnet-typed Li6.75La3Zr1.75Nb0.25O12 nanowires

Jianqi Sun[1], Yaogang Li[1], Qinghong Zhang[1], Chengyi Hou[1], Qiuwei Shi[1, 2], Hongzhi Wang[1]
1.Donghua University
2.Nanyang Technological University

Composite solid polymer electrolytes, integrating advantages (inflammability and electrochemical stability) of inorganic parts and virtues (lower interfacial resistance, flexibility and handy preparation) of polymer matrix, present a promising prospect for safe high-energy solid-state lithium energy storage systems. Herein, one-dimensional garnet-typed Li6.75La3Zr1.75Nb0.25O12 nanowires are successfully synthesized, and the nanowires are complexed with poly(methyl methacrylate), lithium perchlorate to fabricate composite solid polymer electrolyte membranes via a facile solution-casting method. The ionic conductivity is sharply increased from 5.98×10−7 of filler-free solid electrolyte to 2.20×10−5 S cm−1 of composite solid polymer electrolyte at room temperature, because the interaction between nanowires and polymer chains provide more free volume and special ionic conductive channels. In the meantime, the composite solid polymer electrolytes exhibit the wide electrochemical window and good stability against lithium anode. The feasibility of composite solid polymer electrolyte is examined in LiCoO2/Li configuration at 0.2 C under 60 and 70 °C, the cells deliver discharge capacity of 134.6 and 143.6 mAh g−1 at first cycle and kept 92.3%, 71.8% retention after 80 and 65 cycles respectively. The results indicate that the as-prepared composite solid polymer electrolytes will be a promising candidate for solid-state lithium batteries in the future.

**POSTER PRESENTATION**

# Continuous and scalable manufacture of amphibious energy yarns and textiles

Wei Gong
Donghua University

Biomechanical energy harvesting textiles based on nanogenerators that convert mechanical energy into electricity have broad application prospects in the next-generation wearable electronic devices. However, the difficult-to-weave structure, limited flexibility and stretchability, small device size and poor weatherability of conventional nanogenerator-based devices have largely hindered their real-world application. Here, we report a highly stretchable triboelectric yarn that involve unique structure designs based on intrinsically elastic silicone rubber tubes and extrinsically elastic built-in stainless steel yarns. By using a newly developed melt-spinning method, we realize scalable-manufacture of the self-powered yarn. A first ever hundred-meter-length (but not limited to this size) triboelectric yarn is demonstrated. The triboelectric yarn shows a large working strain (200%) and promising output. Moreover, it works superiorly in liquid, therefore shows all-weather durability. We also show that the development of this energy yarn facilitates the manufacture of large-area self-powered textiles and provide a new direction for the study of amphibious wearable technologies.

POSTER PRESENTATION

# Regulation of precursor solution concentration for In-Zn oxide thin film transistors

Yanping Chen
Donghua University

The tunable electronic performance of the solution-processed semiconductor metal oxide is of great significance for the printing electronics. In current work, transparent thin-film transistors (TFTs) with indium-zinc oxide (IZO) were fabricated as active layer by a simple eco-friendly aqueous route. The aqueous precursor solution is composed of water without any other organic additives and the IZO films are amorphous revealed by the X-ray diffraction (XRD). With systematic studies of atomic force microscopy (AFM), X-ray photoemission spectroscopy (XPS) and the semiconductor property characterizations, it was revealed that the electrical performance of the IZO TFTs is dependent on the concentration of precursor solution. As well, the optimum preparation process was obtained. The concentrations induced the regulation of the electronic performance was clearly demonstrated with a proposed mechanism. The results are expected to be beneficial for development of solution-processed metal oxide TFTs.

POSTER PRESENTATION

# Controlling the transformation of intermediate phase under near-room temperature for improving the performance of perovskite solar cells

Xuefei Han[1], Hao Xiong[1], Jiabin Qi[1], Yichuan Rui[2], Xin Zhang[1], Chengyi Hou[1], Yaogang Li[1], Hongzhi Wang[1], Qinghong Zhang[1]
1.Donghua University
2.Shanghai University of Engineering Science

Perovskite crystallizations under different temperature phases have tremendously effects on the performance of perovskite solar cells. Here, we report an effective near-room-temperature fabrication technique to prepare high quality organolead halide perovskite films. By doping a certain amount of cesium cations into precursor solution, smooth and well-crystallized perovskite films with no other impurity phase can be obtained at an annealing temperature of 50 °C. This phenomenon can be interpreted as relatively low energy required for the diffusion of cesium cations into intermediate phase framework, promoting the removal of DMSO and subsequent diffusion of organic cations. Corresponding PSCs has been prepared based on this technique, and the highest PCE of 17.9% has been obtained with annealing temperature of only 50 °C. This work exhibits a well balance between reducing of annealing temperature and increase of device performance and is expected to broaden the range of choices for substrate materials.

POSTER PRESENTATION

# Solvatochromic structural color fabrics with favorable wearability properties

Xinbo Gong
Donghua University

Robust, washable, steam permeable fabrics that display solvatochromic structural color changes are fabricated by a solvent evaporation-driven self-assembly method using poly(styrene–butyl acrylate–acrylic acid) core–shell colloidal microspheres. The colloidal microspheres self-assemble into a periodic arrangement to form a colorless transparent film on fabrics. Because of the lack of refractive index difference between the film and fabric, the fabric retains its original color (black polyester in this work). As soon as the fabric is wetted, its color changes. When the wetted fabric is completely dried, its original black color appears again. This solvatochromic fabrics do not show apparent color fading during repeated stretching or after laundering. The structural color fabrics show promising steam permeability with only 13% decrease compared with original fabrics. Therefore, the wearability of the fabric meets the demands for applications in the textile industry.

POSTER PRESENTATION

# MXene-based flexible $Li^{+}$-capacitors and micro-supercapacitors

Jianmin Li, Hongzhi Wang
Donghua University

MXenes - two-dimensional (2D) transition metal carbides and nitrides - are an emerging class of high-rate pseudocapacitive materials that offer a combination of solution processability, fast surface redox reactions and metallic conductivity. These properties have been exploited to develop various kinds of supercapacitors. In addition, the ability of MXenes to spontaneously intercalate cations, broadens the scope for developing metal-ion capacitors beyond the protic electrolytes. To date, the stable voltage window of MXene-based supercapacitors was limited at 1.5 V in acidic aqueous electrolytes, even for asymmetric devices. We developed a Ti3C2Tx//CDC asymmetric supercapacitor with a voltage window of 2 V, by using neutral electrolyte. What is more, to maximize the electronic conductivity, we developed titanium carbide - poly(3, 4-ethylenedioxythiophene) heterostructures by electrochemical deposition using a non-aqueous monomeric electrolytic bath. The electrochromic microsupercapacitor was carved using an automated scalpel technique. Hybrid microsupercapacitors showed 5-fold areal capacitance and higher rate capabilities (2.4 mF/cm2 at 10 mV/s, retaining 1.4 mF/cm2 at 1000 mV/s) over the pristine MXene microsupercapacitors (455 μF/cm2 at 10 mV/s and 120 μF/cm2 at 1000 mV/s). These studies opened new avenues for developing high-performance multifunctional energy storage devices based on MXene heterostructures.

POSTER PRESENTATION

# Personal thermal management textiles based on large-area graphene papers

Yang Guo, Qinghong Zhang, Yaogang Li, Hongzhi Wang, Chengyi Hou
Donghua University

Lightweight, flexible, and wearing comfortable personal thermal management (PTM) device with desirable performance becomes prevalent because of its potential to wisely adjust our body temperature to a thermally-safe and comfort state. In this work, we report a freestanding, flexible/foldable, and large-area ultrathin graphene papers (GPs) with high thermal conductivity and sensitive electro-thermal response, and their application to wearable bifunctional PTM devices for heating and cooling. The heating part is achieved by taking advantages of its joule heating, while the cooling part benefits from its high thermal conductivity which from the ultrathin and compact lateral structure of the GPs. The promising electrical conductivity grantees the superior Joule heating for extra warmth of 42 °C using a low supply voltage around 3.2 V. Besides, based on its high thermal conductivity, the graphene paper provides passive cooling via thermal transmission from the human body to the environment within 7s. The cooling effect of graphene paper is superior compared with that of the normal cotton fabric, and this advantage will become more prominent with the increased thickness. The present bifunctional graphene paper possesses high durability against bending cycles over 500 times and wash time over 1500 min, suggesting its great potential in wearable PTM. Furthermore, we integrated the GPs into textiles by the techniques of plain weave, co-woven, hollow-out, and kirigami. These wearable technology designs achieve not only the personal thermal management, but also the breathability of the PTM devices. It can bring inspiration to the development of intelligent clothing in the future.

**POSTER PRESENTATION**

# Thermal-shrinking-induced ring-patterned boron nitride wrinkles on carbon fibers

Yinlong Tan, Jia Yan, Biru Hu, Zengyong Chu
National University of Defense Technology

The stabilization of polyacrylonitrile (PAN) fibers is of great significance for the fabrication of high-performance carbon fibers (CFs) and microstructures on the CFs are necessary for some special applications. Here a non-oxygen-approach was developed for the stabilization of PAN fibers using boron trichloride ($BCl_3$) as the crosslinking agent, and ring-patterned boron nitride (BN) surface wrinkles were further demonstrated when the PAN fibers are stabilized under the consecutive gases of $BCl_3$ and ammonia ($NH_3$), due to the thermal-shrinking-induced mismatched strain along the axial direction. Besides, tunable topography including a wrinkle-to-fold transition can be realized via the subsequent high-temperature treatment. After the introduction of ring-like wrinkles on the fibers, the surface wettability of the fiber array can be effectively adjusted to achieve a transition from superhydrophilic to hydrophobic. Therefore, the developed thermal-shrinking method provides a novel strategy for the fabrication of ordered ring-like boron nitride wrinkles on CFs. The obtained fibers are potential candidates for the fabrication of self-cleaning clothes, water collectors, anti-icing coatings, and high performance composites.

**KEYNOTE SPEECH**

# Braided stents for aortic and peripheral artery diseases

Lu Wang, Wen Xue, Fan Zhao, Fujun Wang, Jing Lin, Guoping Guan, Chaojing Li
Donghua University

The artery disease has become one of the most serious health issues in front of people worldwide, especially with human dietary changes and aging population. The mortality caused by cardiovascular diseases ranks the first and they can result in various complications. Percutaneous transluminal angioplasty and stent implantation have been adopted to keep vessels open and maintain blood perfusion. Braided stents, made of helix wires interwoven into cylindrical tubes, are of high flexibility and great applicability. This kind of stents from textile technology is gaining extensive attention nowadays. In this paper, the recent progress in the design and fabrication of braided stents for aortic and peripheral artery disease is summarized and highlighted. It covers the introduction of braiding mechanism, raw materials applied for stents as well as related biomechanical and biocompatible evaluation methods on braided stents. The paper concludes with some perspectives and outlook for braided stents in aortic and peripheral artery disease.

INVITED LECTURE

# Electrospun a composite fibrous membrane for guided bone regeneration

Jidong Li, Shue Jin, Yubao Li
Sichuan University

Guided bone regeneration (GBR) technology is a highly effective treatment performed using a mechanical barrier membrane. The purpose of this study was to fabricate a low-immunogenicity fish collagen (FC) and bioactive nano-hydroxyapatite (n-HA) enhanced poly (lactide-co-glycolide) (PLGA) nanofibrous membrane for guided bone regeneration via electrospinning. The physicochemical properties and biological performance were evaluated to assess the potential of this fibrous composite membrane in GBR application. The results revealed that the interaction between FC and PLGA significantly improved the tensile strength of the PLGA membrane. Moreover, the membranes exhibited favorable cytocompatibility with bone marrow stem cells and human gingiva fibroblasts cells, as well as strong GBR effect in rat skull defect. All the results indicate that this composite fibrous membrane exhibits significant potential for guided bone or tissue regeneration.

INVITED LECTURE

# Hydrogel electrospun fibrous scaffolds for tissue regeneration

Wenguo Cui, Wang Zhen, Wang Jingwen
Ruijin Hospital, Shanghai Jiao Tong University School of Medicine

Electrospinning technology has a wide range of applications in creating extracellular matrix (ECM)-mimicking ultrafine fibrous membranes with fiber diameters ranging from nanometers to micrometers. Many studies reported that 2D- or 3D-structures of electrospinning fibers meshes were fabricated using poly (L-lactic acid) (PLLA) or poly (ε-caprolactone) (PCL) for vascularization. But in vitro results indicated that endothelial cells can only grow on the surface of the fiber mesh due to the small pore size, making it difficult to construct 3D vasculature in the scaffolds. In addition, stiffness material is not suitable for the soft tissue regeneration, which would cause severe inflammatory response in vivo. Distal necrosis of random skin flap is always clinical problematic in plastic surgery. The development of three-dimensional (3D) functional vascular networks is fundamental for the survival of a local random skin flap. Hydrogels as biomaterials have been widely investigated for numerous medical applications. Because of their high water content and elastic similarities between hydrogels and soft tissues in the body, hydrogels are particularly used for tissue engineering, wound healing and as bioadhesives. From the above description, we put forward the concept of electrospinning hydrogel fibers that possess dual properties of electrospun fibrous nanostructure and hydrogel softness, which is expected to allow cell migration into the scaffolds to construct 3D microvascular structures. Hence, soft electrospun hydrogel fibers have potential to be ideal biomaterial scaffold for promoting vascularization. Therefore, we hypothesize that such hydrogel fibrous membranes fabricated by electrospinning are conducive to: (1) endothelial cell adhesion and growth; (2) tubulogenesis; (3) skin flap adhesion of the wound bed; and (4) the formation of microvasculature, which will increase the number of capabilities to aid blood supply, and sequentially enhancing the survival rate of random skin flap after implantation.

**KEYNOTE SPEECH**

# Electrospun fibrous scaffolds for in vitro drug screening and biosensing

Xiaohong Li
Southwest Jiaotong University

Electrospun fibers have off ered several advantages in mimicking the size and morphology of extracellular matrix, and providing a very high fraction of surface available to interact with cells. To mimic the cellular microenvironment, micropatterned fibrous scaffolds are created to establish the coculture of hepatocyte with fibroblasts, and endothelial cells. In order to mimic the myocardial lamellar structure, cardiomyocytes are cocultured on conjunct or sandwiched patterned fibrous mats. The micropatterned cocultured hepatocytes and cardiomyocytes are determined as in vitro model for drug metabolism and drug screening. Additionally, electrospun fibers provide a large surface area, leading to a high availability of graft sites and an increase in the reaction rate for developing high-performance biosensors. In order to overcome the quenching problem of fluorescent probes, a series of aggregation-induced emission chromophores are synthesized and decorated onto electrospun fibers for sensing different targets. To solve the problems that the signal intensity changes may be susceptible to external factors in "turn on" or "turn off" detection, electrospun fibrous strips are developed for ratiometric detection of biomarkers. In order to decrease the detection limit and time, self-amplified electrospun fibrous strips and self-propelled Janus fibers rods are prepared for detection of hydrogen peroxide, bacteria, and circulation tumor cells.

INVITED LECTURE

# Electrospun 3D nanofiber for tissue engineering

Xiumei Mo
Donghua University

Electrospinning nanofiber can biomimetic Extracellular Matrix and suitable for tissue scaffolding. In our group collagen-chitosan complex nanofibers have been fabricated for skin tissue engineering, collagen-chitosan-P(LLA-CL) nanofibers tube scaffold have been fabricated for blood vessel tissue engineering, silk-P(LLA-CL) nanofibers tube scaffold have been fabricated for nerve tissue engineering. Electrospinning fabrication technique most commonly produces relatively 2D mats and the construction 3D structure nanofibers with higher porosity is still a major challenge. In our research, two methods were used to fabricate the 3D nanofiber scaffolds. A dynamic electrospinning method were developed to fabricate the nanoyarn scaffold. The nanoyarn scaffold contained 3D aligned microstructures with larger interconnected pores and higher porosity comparing with nanofiber scaffold. The nanoyarn scaffold have been successfully used for tendon tissue regeneration of rabbit. Gelatin/PLA nanofiberous 3D scaffold was fabricated by using combined electrospinning and freeze-drying methods. Thus obtained 3D nanofiber scaffold could promote cells growth and proliferation, it also been used for cartilage tissue engineering in rabbit.

ORAL PRESENTATION

# 3D printing of biomimetic vasculature for tissue regeneration

Dong Lei1, Yang Yang2, Xiaofeng Ye2, Zhengwei You1
1.State Key Laboratory for Modification of Chemical Fibers and Polymer Materials, International Joint Laboratory for Advanced Fiber and Low-dimension Materials, College of Materials Science and Engineering, Donghua University
2.Department of Cardiac Surgery, Ruijin Hospital, Shanghai Jiaotong University School of Medicine

Mimicking native vascular networks remains a key challenge in tissue engineering. Here we present a simple strategy to fabricate perfusable and permeable hierarchical microchannel-networks (PHMs) via the combination of one-pot 3D printed sacrificial caramel templates and polymer coating with integrated phase seperation[1]. The patterned PHMs possess a biomimetic three level vascular structure including custom-made 3D framework, interconnected microchannels and permeable walls with controllable micropores. The fabrication process can be adapted to various polymers and integrated with diverse matrixes including hydrogels, particle leached porous scaffolds, electrospun nanofibers, and bacterial cellulose. We demonstrated the power of PHMs to facilitate mass exchange in tissue engineering constructs by showing that the PHMs could maintain the metabolic functions of heart cells in vitro, facilitate in vivo angiogenesis and tissue integration, and efficiently treat myocardial infarction.
References: [1] Lei D, Yang Y, Liu Z, et al. 3D printing of biomimetic vasculature for tissue regeneration. Materials Horizons, 2019, 6, 1197-1206

ORAL PRESENTATION

# Physico-chemical and biological evaluation of PLCL/SF nanofibers loaded with oregano essential oil

Atta ur Rehman Khan
Donghua University

Essential oils are complex volatile compounds, extracted from specific plant species, with promising therapeutic potentials. However, their volatile nature presents a major hindrance in using them as therapeutic agents. In the current study, we successfully encapsulated oregano essential oil (OEO) into Poly (l-lactic acid-co-e-caprolactone) /Silk Fibroin (PLCL/SF) polymers through electrospinning. The nanofibrous membrane (NF) was fabricated and characterized for various physico-chemical and biological attributions. Homogenous and bead free morphology was confirmed by scanning electron microscopy (SEM). Attenuated total reflection Fourier transform infrared spectroscopy (ATR-FTIR) confirmed the successful loading of OEO and its physical interaction with the blend of PLCL/SF. Moreover, thermogravimetric analysis (TGA) also confirmed the successful loading and thermostability of the OEO. Although a significant change was noted in tensile strength due to the loading of OEO, the mechanical behaviour still falls into the acceptable ranges required for skin tissue engineering. Similarly, fabricated material was evaluated for its biological significance. Liquid chromatography-mass spectrometry (LC-MS) was employed to determine the release behaviour of OEO from electrospun membranes. LC-MS data, noted for 48 h, confirmed the biphasic release of OEO. Furthermore, NF membranes have shown strong antioxidant and antibacterial properties.This material is promising and can be used for infectious wound healing applications.

**KEYNOTE SPEECH**

# Functional nanogels in the fascinating world of bioengineering

Bhuvanesh Gupta
Indian Institute of Technology

Polymeric biomaterials have become increasingly interesting in medical science over the past few years. With the increasing demand for a biomaterial with better acceptability and functionality to the biosystem, stress has been focused on the development of newer materials. The major requirements for biomedical applications are nontoxicity, the ability to be sterilized, mechanical properties and the most important one is the biocompatibility of the material with the biological systems.

The development of polymers with bioactive coating is an important area of research focused on solving the problem of contamination by infection in wound care systems. We have observed that a bioactive component may be incorporated within the hydrogel matrix to make it infection resistant. Herbal drugs and essential oils have been incorporated into either natural hydrogels like chitosan, pectin and dextran by blending approach. A wide range of natural bioactive agents such as aloe vera, curcumin, sandal wood oil, clove oil and honey are available to develop excellent materials for wound care. These dressings have been evaluated in wound healing using mouse as the animal model. Excellent healing with minimum scar by hydrogel dressings has been observed.

The innovation of present work is to reconstruct the surface of polymers at nano level in such a way that it acquires antimicrobial nature without much altering its bulk properties which are quite appropriate for suture application. The nanoconstruction can be carried out so that surface chemistry of the material is selectively altered. For this we have followed plasma processing as the mean for the surface modification. The most attractive feature of the plasma processing is that by exerting proper control over the exposure conditions such as exposure time, plasma power and gas flow, a tailored surface with desired chemical functionality and morphology may be produced.

References

Anjum S., Gupta A., Sharma D, Shanti, Bora J, Plabita, Bhan S. and Gupta B., 2017, J Bioact Comp Polym., 32, 263

Anjum S., Gupta A., Sharma D, Shanti, Bora J, Plabita, Bhan S. and Gupta B., 2016, J Mater. Sci. Engg., 64, 157

INVITED LECTURE

# Design of Dendrimer-Based CT/MR Contrast Agent for Hypoxic Tumor Precision Imaging and Sensitized Radiotherapy

Xiangyang Shi1, Yu Fan1, Wenzhi Tu2, Yong Liu2
1. Donghua University
2. Shanghai First People's Hospital

Radiation therapy (RT) has been widely used for malignant tumor treatment in the clinic. However, RT efficacy is limited due to the hypoxia inside most of the solid tumors, which are insensitive to RT[1]. Therefore, overcoming the radioresistance caused by tumor hypoxia has been a challenge task for RT. Herein, we present the creation of hypoxia-targeted dendrimer entrapped gold nanoparticles complexed with gadolinium(III) (Gd-Au DENPs-Ni) for dual mode CT/MR imaging of tumor hypoxia and sensitized radiotherapy. In this work, poly(amidoamine) dendrimers of generation 5 were partially decorated with DOTA, then used to entrap Au NPs within their interior, conjugated with nitroimidazole via a PEG linker, followed by chelation with Gd(III) ions via DOTA ligands and acetylation of their remaining amine termini. We reveal that the designed dendrimer-based nanohybrids having an Au core size of 3.2 nm and a hydrodynamic size of 141.4 nm, display a good X-ray attenuation property, reasonable r1 relaxivity (1.32 mM-1s-1), enhanced cellular uptake in hypoxic cancer cells (CNE-1 cells), and enabled effective dual mode CT/MR imaging of subcutaneous CNE-1 xenografted tumor model. Most importantly, the sensitizer enhancement ratio of Gd-Au DENPs-Ni was calculated to be 1.24 via colony formation assay. When irradiated with an X-ray, the Gd-Au DENPs-Ni nanohybrids could produce more reactive oxygen species, promote DNA damage and prevent DNA repair, thereby facilitating the enhanced therapeutic efficiency of RT. As demonstrated by in vivo experiments, the hypoxic tumor was significantly suppressed after the treatment of RT using Gd-Au DENPs-Ni as the nanosensitizer. As a result, this kind of hypoxia-targeted dendrimer-based nanohybrids may be used both contrast agents for dual mode CT/MR imaging of tumor hypoxia and sensitizer for enhanced tumor radiotherapy.

**INVITED LECTURE**

# Tumor microenvironment-responsive dendritic polymers-based nano-agents for cancer treatment

Kui Luo
Huaxi MR Research Center (HMRRC), Department of Radiology, West China Hospital, Sichuan University

Tumor microenvironment (TME)-responsive drug delivery systems that deliver a drug in spatial-, temporal- and dosage-controlled patterns have become a promising way to realized tumor-specific treatment. We moved on to designing a series of biodegradable dendritic polymers for multiple stimuli-responsive drug delivery. These nano-platforms are shown that exhibit good biocompatibility and significant antitumor efficacy. Furthermore, these nano-platforms conjugated doxorubicin, gemcitabine, paclitaxel and oxaliplatin through pH-sensitive hydrazone bond, N, O-coordination and enzyme-sensitive GFLG linker to prepare a series of TME-responsive nanoscale drug delivery systems. Through the systemic delivery of dendronized PEG-platinum (II), 25-fold higher tumor platinum uptake at 36 h post-injection was seen observed due to the enhanced permeability and retention (EPR) effect, which far higher than the reported TME-responsive drug delivery systems. Then we designed prepared a series of TME-responsive branched HPMA nano-platforms via reversible addition-fragmentation chain transfer (RAFT) polymerization, Gd(III) chelating, drug conjugation and cRGDyK functionalization for magnetic resonance imaging and targeted therapy. Finally, we explored systematically the relationship between the structure and their behaviors for drug delivery, and found that amphiphilic dendronized polymers with a moderate HLB value display enhanced stability and highly efficient tumor retention. These high-performance TME-responsive dendritic polymers based nano-platforms may be employed as a safe and efficient multiple stimuli-responsive drug delivery systems for diagnosis and therapy of cancer.

INVITED LECTURE

# Metallodentrimers - a route for new anticancer candidates

J Rodrigues[1, 2*], N Nunes[1], C Camacho[1], D Maciel[1], H Tomás[1]
1. CQM - Centro de Química da Madeira, MMRG
Universidade da Madeira, Campus Universitário da Penteada, 9000-390 Funchal, Portugal.
2. School of Materials Science and Engineering/Center for Nano Energy Materials
Northwestern Polytechnical University

Despite all scientific advances, the search for drugs able of acting as anticancer agents represents one of the primary goals and challenges in research.

Metallocomplexes like Platinum and Ruthenium based compounds have been studied over the last years as anticancer drug candidates, presenting powerful therapeutic properties. Nevertheless, just a few numbers of potential metallodrugs reached the market and are currently in clinical use as anticancer drugs (e.g. Cisplatin, Carboplatin, Oxaliplatin). Several factors can contribute to this situation, but their lower solubility, severe side effects and acquired drug resistance against different types of cancer cells line should be in between of the main reasons[1].

One of the ways to overcome these difficulties, is the incorporation of metallocomplexes onto dendritic scaffolds, producing metallodendrimers[2], that can result in improved anticancer activity, solubility and PK/PD behaviors[3].

Considering the growing interest of metallodendrimers in cancer research and, at the same time, the few number of reports available[4], we herein present our most recent contribution on that domain by using the organometallic moiety $[Ru(\eta5-C5H5)(PPh3)2]+$ or the oxaliplatin complex, cis-dichloro(1, 2-diamminocyclohexane) platinum(II) (DACHPt) functionalised on PAMAM dendrimers.

The prepared metallodendrimers were structurally characterized by NMR, MS, FTIR and DLS. It's in vitro cytotoxicity, hemolytic activity, and reactivity towards DNA was also evaluated during this study.

In general, all the prepared metallodendrimers presenting superior anticancer activity than the parent metalllocomplexes, justifying further biological studies.

[1] K D Mjos, C Orvig, Chem. Rev., 2014, 114, 4540.

[2]P. Govender, B. Therrien, G.S. Smith, Eur. J. Inorg. Chem., 2012, 17, 2853.

[3] S. Mignani, J. Rodrigues, H. Tomas, M. Zablocka, X. Shi, A.-M. Caminade, J.-P. Majoral, Chem. Soc. Rev., 2018, 47, 514.

[4] M. Gouveia, J. Figueira, M.G. Jardim, R. Castro, H. Tomás, K. Rissanen, J. Rodrigues, Molecules, 2018, 23, 1471.

INVITED LECTURE

# Inorganic-organic nanocomposites for the imaging and therapy of tumors

Zhigang Chen
State Key Laboratory for Modification of Chemical Fibers and Polymer Materials, College of Materials Science and Engineering, Donghua University

Near-infrared (NIR) laser-induced photothermal ablation therapy (PAT) has great potential to revolutionize conventional therapeutic approaches for cancers, and a prerequisite is to obtain biocompatible and efficient photothermal nanoagents. To address this problem, we have used hydrophilic polymer as surface ligands to prepare several semiconductor nanoagents, including CuS [1] and $W_{18}O_{49}$ nanomaterials [2]. Their aqueous dispersions exhibit intense NIR absorbance and excellent photothermal effects. Importantly, if we inject aqueous dispersion into the tumor in mice, cancer cells in vivo can be efficiently ablated under the irradiation of NIR laser with a safe intensity for ~10 min. To further improve the therapeutic effects, the combination of NIR-PAT and chemotherapy has been proposed for the synergic therapy. Recently, by combing semiconductor nanoagents and thermosensitive polymer nanogels, we have constructed smart nanocapsules (G-CuS-DOX) that can be switched by NIR-laser[3]. The nanocapsules exhibit the controllable and efficient photothermal/chemotherapy effect compared with single PAT or chemotherapy effect for the tumor. Furthermore, ideal theranostic nanoplatform for tumor should be one nanoparticle that has single semiconductor or metal component but contains all multi-model imaging and therapy abilities. By tuning vacancy concentrations and surface ligands, we have developed $FeS_2$ (size: ~350 nm) [4], Nb-doped $TiO_2$ [5], Sb-doped $SnO_2$ [6], Bi [7] nanoparticle as "all-in-one" multifunctional nanoagents for the imaging guided PAT of tumor. Therefore, these inorganic-organic nanocomposites have great potential in the imaging and/or therapy of tumor.

References

[1] Q. W. Tian, M. H. Tang, Y. G. Sun, R. J. Zou, Z. G. Chen, M. F. Zhu, S. P. Yang, J. L. Wang, J. H. Wang, J. Q. Hu, Adv. Mater. 2011, 23, 3542.

[2] Z. G. Chen, Q. Wang, H. L. Wang, L. S. Zhang, G. S. Song, L. L. Song, J. Q. Hu, H. Z. Wang, J. S. Liu, M. F. Zhu, D. Y. Zhao, Adv. Mater. 2013, 25, 2095.

[3] Z. Meng, F. Wei, R. Wang, M. Xia, Z. Chen, H. Wang, M. Zhu, Adv. Mater. 2016, 28, 245.

[4] Z. Meng, F. Wei, W. Ma, N. Yu, P. Wei, Z. Wang, Y. Tang, Z. Chen, H. Wang, M. Zhu, Adv. Funct. Mater. 2016, 26, 8231.

[5] N. Yu, Y. Hu, X. Wang, G. Liu, Z. Wang, Z. Liu, Q. Tian, M. Zhu, X. Shi, Z. Chen, Nanoscale 2017, 9, 9148.

[6] N. Yu, C. Peng, Z. Wang, Z. Liu, B. Zhu, Z. Yi, M. Zhu, X. Liu, Z. Chen, Nanoscale 2018, 10, 2542.

[7] N. Yu, Z. Wang, J. Zhang, Z. Liu, B. Zhu, J. Yu, M. Zhu, C. Peng, Z. Chen, Biomaterials. 2018, 161, 279.

ORAL PRESENTATION

# Construction and biomedical applications of well-defined porphyrin-based polymers with specific structures

Jia Tian, Baoxuan Huang, Weian Zhang
East China University of Science and Technology

Cancer is a major threat to human health in the world. Recently, photodynamic therapy (PDT) has emerged as a promising cancer treatment due to its merits of spatial and temporal controllability and noninvasiveness.[1–3] Porphyrins are a typical and common photosensitizer in PDT. However, the porphine core of porphyrins is in a planar structure and hydrophobic, which leads to the aggregation of porphyrins, quench of fluorescence (aggregation-caused quench, ACQ), and the reduction of reactive oxygen species (ROS) in PDT.[4] We hypothesized well-defined porphyrin-based polymers with specific structures could significantly increase the solubility in aqueous solution and reduce the aggregation of porphyrins, which further improve the anticancer therapeutic efficacy in PDT.[5]
(1) Block copolymers with porphyrin units as monomers

(2) Block copolymers with a porphyrin unit at the junction point

(3) Polymer brushes with porphyrin units at side chains

Reference
[1] W. Fan, P. Huang, X. Chen, Chem. Soc. Rev. 2016, 45, 6488–6519.
[2] A. Kamkaew, F. Chen, Y. Zhan, R. L. Majewski, W. Cai, ACS Nano 2016, 10, 3918–3935.
[3] B. M. Luby, C. D. Walsh, G. Zheng, Angew. Chemie - Int. Ed. 2019, 58, 2558–2569.
[4] J. Jin, Y. Zhu, Z. Zhang, W. Zhang, Angew. Chemie - Int. Ed. 2018, 57, 16354–16358.
[5] J. Tian, W. Zhang, Prog. Polym. Sci. 2019, 95, 65–117.
[6] J. Tian, B. Huang, H. Li, H. Cao, W. Zhang, Biomacromolecules 2019, 20, 2338–2349.
[7] J. Tian, C. Xiao, B. Huang, X. Jiang, H. Cao, F. Liu, W. Zhang, ACS Appl. Bio Mater. 2019, 2, 1862–1874.

**ORAL PRESENTATION**

# Multi-biofunctional surface constructed by flexible and conductive polypyrrole membranes and functional particles

Jifu Mao[1, 2], Mahmoud Rouabhia[2], Ze Zhang[2], Lu Wang[1]
1.Donghua University
2.Laval University

Introduction: Surface biofunctionalization by immobilization of biomolecules is an important approach to integrate specific bioactivities to materials for biomedical applications1. However, building a multi-biofunctional surface by co-immobilization with controllable and quantitative biomolecules is still a challenge, especially for those biomolecules with significant difference in physical and chemic natures. Meanwhile, lack of biofunctionalization and inferior mechanical properties remain the two principal outstanding obstacles facing pristine polypyrrole (PPy) due to its unique extensive conjugated and cross-linked molecular structure2. In this modular strategy, numerous biomolecules can be covalently grafted onto the PPy particles separately and predetermined the quantity of biomolecules on the particles. After that, diverse modified particles could be mixed and co-immobilized onto PPy flexible membrane according to the requirement, which shows great superiority to co-immobilize more than three kinds of biomolecules.
Materials & Methods: The flexible PPy membranes were synthesized through a template assisted interfacial polymerization (TIP)2. The functional core-shell poly(pyrrole-co-(1-(2-carboxyethyl)pyrrole)) (P(Py-PyCOOH))particles were synthesized through a simple emulsion polymerization3. After that, bovine serum albumin (BSA) and human serum albumin (HSA) as model molecules were covalently immobilized onto the particle surface, separately; and the protein grafted particles were mixed and assembled onto the flexible membranes.
Results & Discussion: The flexible PPy membrane presented an asymmetrical morphology on two sides, i.e., a bubble side and a nanotube side. Frozen drying method confirmed the thickness and morphology change of nanotube side during drying procedure. The thickness of nanotube layer decreased from 175 μm to 14 μm after normal drying. Meanwhile, the porous structure was replaced by packed PPy nanotubes similar to the drying of aquatic plants. Such change offered a great opportunity for the functional particles immobilization. The protein was grafted onto the P(Py-PyCOOH) particles firstly, and then the modified particles were coated onto the flexible PPy membrane. In addition, after the loading of 0.27 mg cm-2 of protein grafted P(Py-PyCOOH) particles, the biofunctionalized PPy membrane remained its outstanding processability and flexibility. This biofunctionalized conductive membrane can be easily cut into different shapes as a paper and rolled into tubes of different diameters. And the epifluorescence microscopy observation confirmed that the PPy membrane loaded with both BSA and HSA grafted particles revealed uniform green and red fluorescent intensity suggesting that two kinds of albumin were successfully.

ORAL PRESENTATION

# Adoptive cellular immunotherapy of tumors via effective CpG delivery to dendritic cells using dendrimer-entrapped gold nanoparticles as a gene vector

Huan Chen, Xueyan Cao, Xiangyang Shi
Donghua University

The major obstacle to hinder current cancer immunotherapy is to develop an effective approach to promote proper immune response for effective tumor killing through activated T cells. Herein, we report an effective T cell-based tumor immunotherapy approach through non-viral delivery of cytosine-guanine (CpG) oligonucleotide using dendrimer-entrapped gold nanoparticles (Au DENPs). In our work, Au DENPs partially decorated with methoxy polyethylene glycol (mPEG) were synthesized and characterized to be used as a vector for CpG delivery to bone marrow-derived dendritic cells (BMDCs). The BMDCs maturated via CpG delivery were used to activate T cells for adoptive immunotherapy of cancer cells. We show that the developed PEGylated Au DENPs are able to effectively transfect CpG to lead the maturation of BMDCs that can be used to activate T cells for subsequent adoptive immunotherapy of cancer cells in vitro and a xenografted melanoma tumor model in vivo after intravenous injection. Importantly, the developed approach to genetically engineering BMDCs enables triggered adaptive immune response and memory of T cells, which may be beneficial for effective inhibition of tumor metastasis and recurrence. The developed nonviral gene delivery approach using Au DENPs as a vector for T cell-based immunotherapy may be applicable for different cancer types.

**KEYNOTE SPEECH**

# Exploring elastomeric biomaterials for multiple medical applications

Biqiong Chen
Queen's University Belfast

Elastomers are of interest to a variety of engineering and medical applications due to their excellent stretchability and resilience. As a result, different types of bioelastomers and their nanocomposites have emerged in the last two decades for various medical applications. At first, biomimetic biodegradable elastomers with controlled mechanical, biodegradation and water absorption behaviours for soft tissue engineering and drug delivery will be presented. Secondly, elastomer nanocomposites with water-activated shape-memory effects will be described, which are promising in minimally invasive medical devices, actuators, etc. Finally, conductive elastomer nanocomposites will be introduced for wearable medical devices for monitoring human motions and health conditions. In all cases, the preparation and characterisation of the elastomeric materials will be described, followed by the evaluation of their potential performance in targeted applications.

INVITED LECTURE

# Bio-inspired polymeric heart valves exhibiting valve-like mechanical and hemodynamic behavior

Xing Zhang
Institute of Metal Research, Chinese Academy of Sciences

Heart valve disease with major symptoms of stenosis and regurgitation is prevalent worldwide. Surgical replacement of diseased heart valves at the end-stages has been widely performed with mechanical valves (MVs) or bioprosthetic heart valves (BHVs). All these current devices have significant limitations with risks of further morbidity and mortality. For example, MVs may cause hemorrhage and thromboembolism, and require anticoagulation for the lifetime of the patients. BHVs show better hemodynamic behavior due to the composition and structural similarity to native heart valves when compared to MVs, however, they do show limited durability because of calcification and progressive degeneration [1]. Thus, polymeric heart valve (PHV) prostheses combining the advantages of MVs and BHVs with long-term durability and no necessity for permanent anticoagulation are of great interest and also show potential applications in advanced transcatheter devices.

In this study, two types of silk fibroin (SF) fiber membranes with anisotropic (ASF) and isotropic (ISF) properties were prepared by electrospinning methods, and were further combined with poly(ethylene glycol) diacrylate (PEGDA) hydrogels to serve as polymeric heart valve (PHV) substitutes (PEGDA-ASF and PEGDA-ISF). The uniaxial tensile tests showed obvious anisotropy of PEGDA-ASF composites with elastic moduli of 10.95 ± 1.09 MPa and 3.55 ± 0.32 MPa, respectively, along the direction parallel and perpendicular to the fiber alignment, close to those of native aortic valve leaflets, while PEGDA-ISF processed isotropic property with elastic moduli of 4.54 ± 0.43 MPa. These novel PHVs consisted of polymeric fibers to mimic the fibrous networks in the fibrosa and ventricularis layers for stress bearing, as well as PEGDA hydrogels to improve anti-fouling function [2, 3]. Furthermore, the presence of PEGDA hydrogels in the composites improved the resistance to progressive calcification of the embedded fibers in vitro, likely due to prevention of large-size hydrated ions to pass through by the polymeric networks of the hydrogels [3]. The non-fouling PEGDA hydrogels encapsulated the surfaces of the composites and prevented contact between platelets and the underlying fibers [4].

Pulse duplicator tests presented good hydrodynamic characteristics of these PHVs from PEGDA-ASF and PEGDA-ISF composites according to the ISO 5840-3 standard. Finite element analysis (FEA) revealed the PEGD-ISF valve with anisotropic property showed a lower peak maximum principle stress value (2.20 MPa) in commissures during diastole compared to that from the isotropic PEGD-ISF valve (2.37 MPa). In systole, the bending area of the PEGDA-ISF valve was close to free edges, however, which appeared in the belly portion and near the attachment line for the PEGDA-ASF valve. Hence, our results revealed that anisotropic properties played important roles not only in mechanical properties, but also in hydrodynamic performance of these artificial PHVs. These novel PHVs with good biocompatibility and hemodynamic property can likely be used for heart valve replacement in future.

INVITED LECTURE

# Construction of porphyrin-containing well-defined polymers and their application in photodynamic therapy

Weian Zhang
East China University of Science and Technology

Photodynamic therapy (PDT) has recently emerged as a promising noninvasive and safe treatment protocol for cancers and other non-malignant diseases. PDT involves the site-specific delivery of a photosensitizer and subsequent singlet oxygen production upon photo-irradiation in the presence of oxygen, leading to the irreversible cytotoxicity to the tumor tissue. Porphyrin and its derivatives as some of the most important photosensitizers for PDT have been widely studied and applied in clinic. However, limitations still exist in treatments with porphyrinic photosensitizers, which exhibit low hydrophilicity, poor biocompatibility, and non-selectivity. To overcome these barriers, in the past several years, we constructed a series of porphyrin-containing well-defined polymers using porphyrins as initiators, monomers, RAFT transfer agents and "click chemistry" agents. A variety of assemblies based on porphyrin-containing amphiphilic polymers have been fabricated for PDT, where assemblies could effectively improve the properties of porphyrinic photosensitizers and enhance the efficacy of PDT.

INVITED LECTURE

# Bioelastomers, 3D printing and their diverse applications

Zhengwei You
Donghua University

Biodegradable and biocompatible elastomers (bioelastomers) especially thermoset ones have biomimetic mechanical properties and attractive for biomedical applications.[1] 3D printing of bioelastomers offers great potential for advanced and precise biomedical applications, however, remains a challenge due the long term curing of thermosets. Herein, we present a versatile strategy to direct 3D print various thermosets using sacrificial salt particles as the removable thicker for printing and reinforcer for curing exemplified by crosslinked polyester, polyurethane and epoxy resin.[2] Specifically, 3D constructs exhibited hierarchical and porous architectures, highly desired for many applications including regenerative medicine, sensors, catalyst supports, and energy absorbers. To demonstrate proof-of-concept, we 3D printed poly(glycerol sebacate) bioelastomer for vapomechanical sensors and actuators featuring fast, robust, and gradient responsiveness, biodegradable elastic myocardial patches to efficiently treat myocardial infarction, and customized sustainable self-powdered wearable electronics based on triboelectric nanogenerators.[2-3] Furthermore, combining with newly designed 4-axis printing technology, we fabricated porous PGS tubular scaffold to from tubular mature cartilage-like tissue with a great potential for tracheal cartilage reconstruction.[4] Moreover, the carrier material (salt) can be easily replaced by or combined with other additive materials, such as nano-clays, graphene and carbon nanotubes, to fabricate high performance and functional thermosetting composites. In addition, we will present a simple versatile strategy to fabricate perfusable and permeable biomimetic vasculature via the combination of one-pot 3D printed sacrificial caramel templates and polymer coating including bioelastomers with integrated phase separation.[5] This technology will be very useful for in vitro disease model, in vivo tissue regeneration, and construction of 3D microfluidics. We expect these studies will pave new ways to tailor the sophisticated 3D structures of polymers, enabling myriad novel applications.

ORAL PRESENTATION

# Urethra-inspired biomimetic scaffold: A therapeutic strategy to promote angiogenesis for urethral regeneration in a rabbit model

Baoxiu Wang1, Xiangguo Lv2, Zhe Li3, Minghao Zhang1, Jingjing Yao1, Nan Sheng1, Mujun Lu2, Huaping Wang1, Shiyan Chen1
1.State Key Laboratory for Modification of Chemical Fibers and Polymer Materials, College of Materials Science and Engineering, Donghua University
2.Department of Urology and Andrology, Shanghai Renji Hospital, Shanghai Jiao Tong University
3.Jiaxing University

Limited angiogenesis and epithelialization makes urethra regeneration using conventional tissue-engineered grafts a great challenge. Consequently, inspired from the native urethra, bacterial cellulose (BC) and bladder acellular matrix (BAM) were designed as a three dimensional (3D) biomimetic scaffold. The developed BC/BAM scaffold was engineered for accelerating urethra regeneration by enhancing angiogenesis and epithelialization. The BC/BAM scaffold reveals the closest mimic of native urethra in terms of the 3D porous nanofiber structure and component including collagen, glycosaminoglycans, and intrinsic vascular endothelial growth factor (VEGF). In vitro studies showed that the bioinspired BC/BAM scaffold promoted in vitro angiogenesis by facilitating human umbilical vein endothelial cells (HUVECs) growth, VEGF secretion, expression of endothelial function related proteins and capillary-like tube formation. Effect of the BC/BAM scaffold on angiogenesis and epithelialization was studied by its implantation in in a rabbit urethral defect model for 1 and 3 months. Results demonstrated that the improved blood vessels formation in the urethra-inspired BC/BAM scaffold significantly promoted epithelialization and accelerated urethra regeneration. The urethra-inspired BC/BAM scaffold provides us a new design approach to construct grafts for urethra regeneration.

**ORAL PRESENTATION**

# Preparation and evaluation of mercerized bacterial nano-cellulose artificial blood vessels

Gaoquan Hu1, 2, FengF Hong1, 2
1.College of Chemistry, Chemical Engineering and Biotechnology, Donghua University
2.Scientific Research Base of Bacterial Nanofiber Manufacturing and Composite Technology, China Textile Engineering Society, China

Bacterial nanocellulose (BNC) is a kind of natural polysaccharides mainly synthesized by acetic acid bacteria and it has many excellent properties such as high porosity and nanofiber structure, etc. Small caliber (<6 mm) BNC tubes have been synthesized by some special bioreactors in our previous work[1], but the application of the BNC tubes in vascular grafts is limited by its weak mechanical properties, the lack of compliance and the thick tube wall that cannot match the natural tissue vessels. It is therefore necessary to modify the structure and enhance the mechanical properties of the BNC tubes to improve the success rate of transplantation. Mercerization is an approach by using the process of alkaline treatment to change the structure of natural fibers. Undergone the alkaline treatment of plant-based cellulose, cellulose Ⅰ can be transferred to cellulose Ⅱ and different physicochemical properties are given. In this research, BNC tubes were mercerized by the alkaline treatment to endow them better mechanical properties and controllable structures[2]. The influence of reaction time, temperature and the concentration of NaOH solution on mechanical properties of the mercerized BNC tubes was investigated and their potential to be artificial blood grafts was also evaluated by assessing the hemocompatibility, cytocompatibility and biocompatibility. This research was aimed to obtain the mercerized small caliber BNC vascular grafts with excellent mechanical properties and biocompatibility as well as compatible wall thickness to lay a better foundation for its applications.

References:
[1] TANG J, LI X, BAO L, et al. Comparison of two types of bioreactors for synthesis of bacterial nanocellulose tubes as potential medical prostheses including artificial blood vessels [J]. Journal of Chemical Technology & Biotechnology, 2017, 92(6): 1218-1228.
[2] HONG F, BAO L, TANG J, et al. Alkali shrinkage bacterial nanocellulose tube as well as preparation method and application thereof. CN20191093212, 20190130

**KEYNOTE SPEECH**

# 仿生手性水凝胶

冯传良
上海交通大学

水凝胶是最好的仿生细胞外基质（ECM）之一，而 ECM 是支持并连接组织的网架结构，主要调节组织的发生和细胞的生理活动，其中 ECM 手性特征起到了决定作用，然而关于 ECM 手性调节组织细胞的研究仍旧处于探索阶段，其影响细胞粘附、生长的本质和机理仍旧是个亟需解决的关键科学问题。近几年来，我们课题组将可功能化的理念引入到了超分子凝胶分子结构的设计中，合成了系列基于 C2 对称的手性超分子凝胶因子，具有模块可设计的特点，构建了系列手性可控、功能可调、智能响应的仿生纳米纤维自组装体，水分含量最高可达 99.94%，拥有 10-80 微米多孔结构，纳米纤维直径在 10-300 纳米之间，满足了细胞生长空间的需求。围绕手性水凝胶材料的基础和应用研究，提出双手性中心凝胶基元的设计思路，发现了长程传递手性的水凝胶构筑原理和方法，赋予了水凝胶手性可调控的特征，揭示了细胞与手性纳米结构相互作用具有选择性的规律，发展了手性结构在组织工程学中重要性作用的理论基础，探索了手性水凝胶多重功能的特征，指导了以手性水凝胶为基础的细胞外培养基质的应用开发，为基于手性仿生微环境的组织工程学研究提供了新思路和方法。

INVITED LECTURE

# The biological oxidation inspired bionmedical hydrogels

Qigang Wang
Tongji University

Due to its 3D crosslinked networks and adjustable physicochemical properties, hydrogels have been widely applied in tissue engineering, drug-delivery system, pollution regulation, polymer electrolyte, agricultural drought-resistance, cosmetic and food area. However, the harsh prepared conditions and high chemical residues of traditional hydrogel both seriously limited their bio-related applications. We introduced the recent advances on tandem enzyme complex for the radical polymerization and further 3D bio-printing of hydrogel.1, 2 We also demonstrated that substrate channel in hydrogel networks can benefit the superactivity of the laden cascade enzymes, which can realize the industrial application of biological oxidation. Finally, we fabricated an enzyme-laden nanogel to mimic neutrophils for antimuor via cascade enzymatic elevation of ROS. This novel biochemical approach is highly useful for cancer therapy. 3, 4, 5

References:
[1] Wang Xia, Chen Shuangshuang, Wu Dongbei., Wu Qing, Wei Qingcong, He Bin*, Lu Qinghua *, Wang Qigang*. Oxidoreductase Initiated Radical Polymerizations to Design Hydrogel and Micro/Nanogel: Mechanism, Molding, and Applications. Advanced Materials 2018, 30, 1705668.
[2] Wei Qingcong, Xu Mengchi, Liao Chuanan, Wu Qing, Liu Mingyu, Zhang Ye*, Wu Chengtie*, Cheng Liming, Wang Qigang*. Printable hybrid hydrogel by dual enzymatic polymerization with superactivity. Chemical Science 2016, 7, 2748-2752.
[3] Wang Xia, Yu Xiao, Wang Xiaoshan, Qi Meiyuan, Pan Jing, Wang Qigang*. One-step Nanosurface Self-assembly of D-peptides Renders Bubble-Free Ultrasound Theranostics. Nano Letters 2019, 19, DOI: 10.1021/acs.nanolett.8b04632.
[4] Wang Xia, Niu Dechao, Li Pei, Wu Qing, Bo Xiaowan, Liu Boji, Bao Song, Su Teng, Xu Huixiong*, Wang Qigang*. Dual-Enzyme-Loaded Multifunctional Hybrid Nanogel System for Pathological Responsive Ultrasound Imaging and T-2-Weighted Magnetic Resonance Imaging. ACS Nano 2015, 9, 5646-5656.
[5] Wu Qing, He Zhigang, Wang Xia, Zhang Qi, Wei Qingcong, Ma Sunqiang, Ma Cheng, Li Jiyu*, Wang Qigang*. Cascade enzymes within self-assembled hybrid nanogel mimicked neutrophil lysosomes for singlet oxygen elevated cancer therapy. Nature Communications 2019, 10, 240.

**INVITED LECTURE**

# Host defense peptide mimicking antimicrobial polymers in solution and on surfaces

Runhui Liu, Yuxin Qian, Qiang Zhang
East China Univeristy of Science and Technology (ECUST)

Drug-resistant microbial infection has been a major challenge to human health. Therefore, considerable effort has been devoted to discovery of new types of antimicrobial agents to battel with drug-resistant microbes. Natural Host defense peptides (HDP) display broad-spectrum antimicrobial activity and low possibility for microbes to develop resistance. However, HDP normally only have moderate activity against bacteria, low stability due to hydrolysis by protease, and high price in preparation. We studied nylon-3 polymers (β-peptide polymers) as mimics of HDP to address above limitations of HDP. These nylon-3 polymers display potent activity against multiple drug-resistant bacterial species, including methicillin-resistant S. aurues (MRSA). These polymers also have low hemolysis and cytotoxicity. Moreover, after repeated use of these antimicrobial β-peptide polymers, bacteria didn't acquire resistance.
When these β-peptide polymers are tethered on the surfaces, selected polymers can still display efficient killing on bacteria including MRSA. The polymer-modified surfaces showed negligible hemolysis and toxicity. Moreover, the polymer-modified surfaces supported the mammalian cell adhesion and fast growth on the surface, which imply application of these polymers as potential biomaterials with good biocompatibility.

References

Wu, Y.; Zhang, D.; Ma, P.; Zhou, R.; Hua, L.; Liu, R.* Lithium hexamethyldisilazide initiated superfast ring opening polymerization of alpha-amino acid N-carboxyanhydrides. Nature Communications 2018, 9 (1), 5297.
Ding, X.; Duan, S.; Ding, X.; Liu, R.*; Xu, F.* Versatile Antibacterial Materials: An Emerging Arsenal for Combatting Bacterial Pathogens. Advanced Functional Materials 2018, 28, 1802140.
Qian, Y.; Qi, F.; Chen, Q.; Zhang, Q.; Qiao, Z.; Zhang, S.; Wei, T.; Yu, Q.; Yu, S.; Mao, Z.; Gao, C.; Ding, Y.; Cheng, Y.; Jin, C.; Xie, H.; Liu, R.*, Surface Modified with a Host Defense Peptide-Mimicking beta-Peptide Polymer Kills Bacteria on Contact with High Efficacy. ACS Appl. Mater. Interfaces 2018, 10 (18), 15395-15400.

ORAL PRESENTATION

# A biodegradable functional water-responsive shape memory polymer for biomedical applications

Yifan Guo, Ziying Lv, Zhengwei You
Donghua University

Shape memory polymers (SMPs) have exhibited great potential in biomedical applications. However, the typical triggers of shape recovery such as heat, UV light, and electricity may be harmful to humans. Accordingly, water-responsive SMPs have become significant, especially for in vivo applications, due to the intrinsic biocompatibility and ready availability of water. However, the reported water-responsive SMPs are limited and relatively complicated. Here, we design a new water-responsive SMP, poly(butanetetrol fumarate) (PBF); the properties of PBF could be modulated by curing. The cured PBF scaffolds exhibited high shape recovery and fixity rates (495%). PBF showed good biodegradability, and it could support the attachment, viability and alkaline phosphatase activity of osteoblasts. Furthermore, PBF could be readily functionalized via pendant hydroxyl groups, which was demonstrated by the immobilization and controlled release of bone morphogenetic protein 2. We expect that PBF will be useful for various biomedical applications including water-responsive scaffolds, sensors or actuators.

References

1. Yifan Guo, Ziying Lv, Yiru Huo, Lijie Sun, Shuo Chen, Zenghe Liu, Chuanglong He, Xiaoping Bi, Xianqun Fan and Zhengwei You.* A biodegradable functional water-responsive shape memory polymer for biomedical applications. J. Mater. Chem. B, 2019, 7, 123--132.
2. Xie, Y.; Lei, D.; Wang, S.; Liu, Z.; Sun, L.; Zhang, J.; Qing, F-L.; He, C.*; You. Z.* A biocompatible, biodegradable and functionalizable copolyester and its application in water-responsive shape memory scaffold ACS Biomater. Sci. Eng. 2019, 5(4) 1668-1676
3. You, Z.; Wang, Y.* A versatile synthetic platform for a wide range of functionalized biomaterials. Adv. Funct. Mater. 2012, 22, 2812-2820.
4. You, Z.; Cao, H.; Gao, J.; Paul, S. H.; Day, B. W.; Wang, Y.* A functionalizable polyester with free hydroxyl groups and tunable physiochemical and biological properties. Biomaterials 2010, 31, 3129-3138.

**ORAL PRESENTATION**

# Development a multi-functional wound dressing using bacterial nanocellulose

Mina Shahriari khalaji1, 2, Feng Hong1, 2
1.Group of Microbiological Engineering and Industrial Biotechnology, College of Chemistry, Chemical Engineering and Biotechnology, Donghua University
2. Scientific Research Base of Bacterial Nanofiber Manufacturing and Composite Technology, China Textile Engineering Society

Bacterial nanocellulose (BNC) and its composites promote major advances to the current and future biomedical materials[1]. BNC is a natural biopolymer with unique properties in particular its nanofiber network, but the general lack of antibacterial activity and wound healing capability limits its superior application as a functional wound dressing[2, 3]. The present study tried to overcome the drawbacks, BNC was therefore incorporated with sodium alginate (SA), chitosan (CS), and silver nitrate ($AgNO_3$).
BNC membrane was immersed into SA solution, then BNC/SA were immersed into a mixture of CS and $AgNO_3$ at different concentrations. The obtained wound dressing material was then examined by using scanning electron microscope, swelling ratio and whole blood clotting. CCK-8 and live/dead fluorescence staining were used to evaluate the cytotoxicity and cell viability of the synthesized material against living cells L929 mouse fibroblasts.
The obtained BNC/SA/CS-Ag material exhibited good antibacterial activity against both gram-positive bacterium Staphylococcus aureus and gram-negative bacterium Escherichia coli. Cytotoxicity test showed that the new dressing material was not toxic to L929 cell line, and the live/dead staining exhibited the majority of L929 cells were alive. All of the results showed that the synthesized wound dressing material had great potential in application as an antimicrobial functional wound dressing. In vivo test will be performed using a rat model in the near future to evaluate the performance of the composite.

1. Wei, B., G. Yang, and F. Hong, Preparation and evaluation of a kind of bacterial cellulose dry films with antibacterial properties. Carbohydrate Polymers, 2011. 84(1): p. 533-538.
2. Yang, G., et al., Preparation and characterization of BC/PAM-AgNPs nanocomposites for antibacterial applications. Carbohydrate Polymers, 2015. 115: p. 636-642.
3. Zhang, P., et al., Using In situ dynamic cultures to rapidly biofabricate fabric-reinforced composites of chitosan/bacterial nanocellulose for antibacterial wound dressings. Frontiers in microbiology, 2016. 7: p. 260-260.

INVITED LECTURE

# Amino acid based colymers for biomedical applications

Jun Wu
School of Biomedical Engineering, Sun Yat-sen University

Poor drug loading capacity and low tumor accumulation of current delivery systems are the main obstacles that prevent nanomedicine from maximizing therapeutic effect. Aiming to achieve high drug loading and excellent tumor accumulation, an amino acid based polymer library was developed via condensation method from natural amino acids, fatty diols and fatty diacids. The structure-function relationship of the polymers was systematically investigated and the results indicated that the polymer structure may have profound effect on physicochemical properties of polymers (such as hydrophobicity). The drug (small molecule drugs, protein/peptides and nucleic acids) loading capability of formulated nanoparticles could be determined by the hydrophobicity and charge property of polymers. Furthermore, the polymers were fabricated into 3-D scaffolds for cell culture and the results showed that the amino acid based polymers could be ideal scaffold materials.

**ORAL PRESENTATION**

# An effective platform for study and discover functional peptides in regulating cell behaviors

Qi Chen, Runhui Liu
East China University of Science and Technology

By modifying bioactive molecules (e.g., cell adhesion peptides (CAPs)) on material surfaces, controlling cell-material surface interactions effectively is essential to the development of improved and new biomaterials for in vitro and in vivo applications such as implantable devices and regenerative medicine. However, only a handful of CAPs are identified and their functions are not fully understood. To identify new CAPs and explore their genuine interactions with specific cells, it is critical to decouple the nonspecific cell adhesion due to biofouling that is unfortunately universal unless antifouling strategies are applied. Polyethylene glycol (PEG) has been extensively used to provide the antifouling layer for covalent immobilization of CAPs. Nevertheless, the impact of PEG antifouling layer on CAPs' function is very less studies. Therefore, we provide a general guideline in selecting optimal PEG molecules for the identification, study, and application of CAPs and other functional peptides, which helps to avoid confusing and even controversial results, and fully display the genuine function of CAPs and other functional peptides.

References

Chen, Q.; Yu, S.; Zhang, D.; Zhang, W.; Zhang, H.; Zou, J.; Mao, Z.; Yuan, Y; Gao, C.; Liu, R.*, Impact of Antifouling PEG Layer on the Performance of Functional Peptides in Regulating Cell Behaviors. J. Am. Chem. Soc. DOI: 10.1021/jacs.9b07105

ORAL PRESENTATION

# A biodegradable antibacterial nanocomposite based on oxidized bacterial nanocellulose for fast hemostasis and wound healing

Haibin Yuan1, 2, Lin Chen1, Feng Hong1, 2, 3
1.Group of Microbiological Engineering and Industrial Biotechnology, College of Chemistry, Chemical Engineering and Biotechnology, Donghua University
2.Key Lab of Science and Technology of Eco-textile, Ministry of Education, Donghua University
3.Scientific Research Base of Bacterial Nanofiber Manufacturing and Composite Technology

Developing biodegradable and antibacterial hemostatic materials with high blood absorption to cease the internal hemorrhage of deep non-compressible wounds remains a challenge. In this study, a novel hemostatic nanocomposite (OBC/COL/CS) was first fabricated by coupling oxidized bacterial cellulose (OBC) and chitosan (CS) to integrate collagen (COL). In other words, during the electrostatic self-assembly process of OBC/CS, COL was cleverly loaded as a functional factor under the electrostatic attraction of cationic CS and anionic OBC. The introduction of collagen is expected to expend the functions such as enhanced hemostasis and promoted wound healing to achieve a new functional composite. This study is the first evaluation on the performance of the OBC, the OBC/CS and the OBC/COL/CS composite for fast internal hemostasis by using a rat liver injury model. To our knowledge, it is also the first to report OBC has a better in vivo biodegradability than the commercial hemostatic oxidized regenerated plant cellulose (ORC). The OBC/COL/CS nanocomposite exhibited proper mechanical strength, broad spectrum antimicrobial property and remarkable degradation in vivo. Moreover, the notable hemostatic efficacy of the composite was confirmed in vivo. The OBC/COL/CS exhibited better blood-clotting ability, higher adhesion of blood cells and platelet, lower blood loss, as well as ultrafast bleeding cessation, which is superior to the commercial hemostatic ORC product SurgicelTM gauze. All the results suggest that the OBC/COL/CS is a fast and efficient procoagulant agent with good antibacterial activity and has great potential to serve as absorbable hemostats for internal bleeding control.

**KEYNOTE SPEECH**

# Smart skin-like materials

Peiyi Wu
Donghua University

Intrinsically stretchable conductors have undergone rapid development in the past few years and a variety of strategies have been established to improve their electro-mechanical properties. However, ranging from electronically to ionically conductive materials, they are usually vulnerable either to large deformation or in harsh environments, mainly due to the fact that conductive domains are generally incompatible with neighboring elastic networks. Herein, we introduce synergistic effect between conductive nanochannels created by zwitterionic polymers and dynamic networks to break the limitations. The as-prepared conductor is highly transparent, ultra-stretchable, with high-strength, self-healing, and capable of maintaining stable conductivity during large deformation and in harsh environments. Transparent integrated sensory systems are further demonstrated via 3D printing of its precursor and could achieve diverse sensations simultaneously towards strain, temperature, humidity, etc., and even recognition of different liquids. This work may break the electro-mechanical limitations encountered with current stretchable conductors, opens up a horizon for the rational designs of their nanostructures, and could promote the development of next-generation of wearable sensory systems for entirely soft robots which may require high transparence, large deformation, and environmental stability.

INVITED LECTURE

# Smart PEDOTs to Selectively Electro-Couple Cells

Bo Zhu1, Hsing-An Lin1, Qichao Pan2, 1, Sihao Qian2, 1, Yaqiong Zhang2, 1
1.Shanghai University
2.Donghua University

Advances of bioelectronics towards high-resolution communication and long-term implantation are reaching limitations of traditional conductive and insulating materials. An ideal bioelectronic material should combine softness, protein-resistance and cell-targeted electro-coupling to meet the required biocompatibility and electrical trade-offs for interfacing with cells/tissues. Toward this aim, we have synthesized a series of biomimicking EDOT polymers with a static, dynamic or 3D specific interaction of high selectivity. All these conductive polymers have demonstrated their superior performance to ensure an intimate, stable and efficient electrical electro-coupling with targeted cells, which are achieved by integrating nonspecific-binding resistance, specific interaction and low-impedance on one conducting polymer.

ORAL PRESENTATION

# Biomedical applications using entropic wetting and thermodynamics of tissues

Ping Wu
Singapore University of Technology and Design

Water and tissue are the main constituents in human body; physical chemistry of water/tissue interfaces is, therefore, an important subject in biomedical applications such as bio-imaging and drug delivery. Although of heavy investigations on the fundamental principles that govern water wetting process and tissue structure stability, these studies have yet to deliver direct design and realistic solutions to biomedical practices. To address this urgent need, the speaker will present; (a) a new water wetting theory based on system entropy, and (b) an intracranial aneurysm rupture risk assessment based on tissue thermodynamics, which were published recently (1, 2).

We do not use the conventional Young's equation and the surface/interface stress concepts, but established a new principle to calculate and explain wetting process from system entropy (1). In this work, contact angle (θ25) at room temperature of a water droplet on a weakly-reactive surface is directly calculated from an entropic equation: S25 = 335.86 – 0.65θ25 (J. unit-mass-1.k-1), where S25 is the entropy of the air/water/substrate system. This entropic equation can be used to advance a wide range of applications from, protein adsorption, blood compatibility, to cell-surface interactions. Our study also opened a new field in physics to understand and characterize contact angles from state entropy.

Furthermore, we demonstrated how to calculate tissue structure stability based on thermodynamic principles. To assess intracranial aneurysm (IA) rupture risk from patient's IA and internal carotid artery (ICA) angiography images, we established three equations from tissue thermodynamics. We aim to: (1) establish foundations to help understand the physics behind the observed morphological changes prior to IA rupture, and (2) provide first-principles equations to aide in rupture risk assessment. These equations are further validated by using available experimental and numerical simulation results. New IA rupture physics are revealed; among the three common structure failure modes, the axial stress and the bending moment loading are the control mechanism.

Overall, the speaker presents a new paradigm on water/tissue interface science based on thermodynamics, which may ignite new ideas, practice and innovations in biomedical applications.

**ORAL PRESENTATION**

# Biofunctionalized chondrogenic shape-memory ternary scaffolds for efficient cell-free cartilage regeneration

Huixia Xuan1, Haoran Hu2, Zhengwei You1, Shichang Zhao2
1.Donghua University
2.Shanghai Jiao Tong University

Cartilage defect repair remains a great clinical challenge due to the limited self-regeneration capacity of cartilage tissue. Surgical treatment of injured cartilage is rather difficult due to the narrow space in the articular cavity and irregular defect area. Herein, we designed and fabricated chondrogenic and physiological-temperature-triggered shape-memory ternary scaffolds for cell-free cartilage repair, where the poly (glycerol sebacate) (PGS) networks ensured elasticity and shape recovery, crystallized poly (1, 3-propylene sebacate) (PPS) acted as switchable phase, and immobilized bioactive kartogenin (KGN) endowed the scaffolds with chondrogenic capacity. The resultant scaffolds exhibited shape-memory properties with shape-memory fixed ratio of 97.6% and recovered ratio of 96.8% at 37 °C for PPS/PGS/KGN-100, indicating a good potential for minimally invasive implantation. The scaffolds gradually degraded in Dulbecco's phosphate-buffered saline and released KGN up to 12 weeks in vitro. In addition, the scaffolds promoted chondrogenic differentiation while inhibiting osteogenic differentiation of bone marrow-derived mesenchymal stem cells in a concentration-dependent manner and cartilage regeneration in full-thickness defects of rat femoropatellar groove for 12 weeks. Consequently, the PPS/PGS/KGN-100 scaffolds efficiently stimulated the formation of an overlying layer of hyaline cartilage mimicking the characteristic architecture of native articular cartilage even in the absence of exogenous growth factors and seeded cells. This study provides much inspiration for future research on cartilage tissue engineering.

ORAL PRESENTATION

# Colorimetric assay and antibacterial activity of wool fabrics treated with Alum/m-TGase and dyed with madder

Reza Assefi Pour1, Jinxin He1
1.College of Chemistry, Chemical Engineering & Biotechnology
2. Chemical Engineering & Biotechnology

This study examined the antibacterial activity of wool fabric treated with Alum/m-TGase and dyed with Madder, the root of Rubia tinctorum L., Alizarin which is one of the oldest natural dyes. To do this, wool fabric was mordanted with alum, treated with microbial transglutaminase (m-TGase) and then dyed with madder. It is well-known that antimicrobial activity of textiles is very important due to different health problems so that recently using antibacterial textiles has received more attention. Since woolen fabric is a natural fiber when dyed with natural dyes can be used in especial industry such as hand-made carpets. Interestingly, making antibacterial carpet can prevent the entrance of many microbes to home. In this regard, herein we tried to find out the optimum condition to achieve the best color after dyeing the wool fabrics with aqueous extract of madder as well as considering the antibacterial activity. Different concentration of Alum and m-TGase were used and antibacterial behavior of dyed wool fabric with madder were investigated against Staphylococcus aureus, SA, as gram-positive bacteria. The investigation of antibacterial activity of dyed wool, treated wool with alum as well as treated wool with Alum/m-TGase and dyed with Madder (different concentrations) indicated that by adding suitable amount of alum and m-TGase, more antibacterial activities could be found. Moreover, the samples were assessed for color strength (K/S), and color fastness and the best conditions of dyeing process are discussed in this study. Totally investigations in color performances of dyed substrate with various composition of Alum and m-TGase demonstrated the enhancement in color fastness as well as color strength and antibacterial activity at optimum condition.

**POSTER PRESENTATION**

# Polyelectrolytes-modified biocomposite scaffold for bone tissue engineering application

Xiaojun Zhou1, 2, JInwu Wang2, Chuanglong He1
1.Donghua University
2.Shanghai Ninth People's Hospital, Shanghai Jiao Tong University School of Medicine

For bone tissue engineering application, a porous scaffold has numerous advantages compared with a dense one, which could offer proper oxygen/nutrient delivery, tissue ingrowth as well as neovascularization. To improve the bioactivity of polymeric sacffold, functional factors were incorporated. Additionally, the low degree of surface hydrophilicity and functional groups will be likely a serious obstacle to further biomedical application for structurally optimized scaffolds. Herein, the polymeric scaffold with macroporous and nanofibrous structure that composed of poly(L-lactic acid) (PLLA), poly(lactic-co-glycolic acid) (PLGA) and polycaprolactone (PCL) were fabricated using thermally induced phase separation (TIPS) technique. The biocomposite scaffold containing strontium-substituted hydroxyapatite (SrHA) was also prepared. Then layer-by-layer assembly of polyelectrolytes (chitosan/gelatin) on porous scaffold was performed to improve the hydrophilicity and protein adsorption capacity so that to construct desirable bone implants. The pristine scaffold exhibited interconnected macropores and biomimetic nanofibrous structure. Water infiltrated quickly into polyelectrolytes-coated scaffold, while no change was observed on pristine scaffold, suggesting good hydrophilicity of the polyelectrolytes-modified scaffold. The polyelectrolytes-modified scaffold also displayed better protein adsorption capacity. Importantly, the biocomposite scaffold exhibit excellent osteogenic capacity, as demonstrated by in vitro and in vivo experiments. Therefore, the constructed biocomposite scaffold may be highly promising as local implantable scaffolds for potential applications in bone tissue engineering.

POSTER PRESENTATION

# Gas foamed three-dimensional electrospinning scaffold for cell penetration

Yujie Chen1, Xianrui Xie1, Xianghao Xiao1, Jinglei Wu1, Rodrigues João2, Castro Rita2, Xiumei Mo1
1.Collage of Chemistry, Chemical Engineering and Biotechnology, Donghua University, China
2.CQM- Centro de Química da Madeira, MMRG, Universidade da Madeira, Campus Universitário da Penteada, 9020-105 Funchal, Portugal

Electrospun nanofibers mimic the original structure of the extracellular matrix (ECM) and are widely used in scaffold materials for tissue repair and regeneration [1, 2].Traditional nanofiber membranes have only a porous structure on the surface. The pore size between the fibers is usually smaller than the size of a single cell, limiting the transport of oxygen and nutrients, making it difficult for cells to penetrate and grow.To increase the pore size of electrospun nanofiber scaffolds, a simple and straightforward way is to modulate fiber diameters.The issue for this method is that the fibers with a size in micron scale lack the biomimetic property.

Here, we prepared P(LLA-CL) with silk fibroin (SF) nanofiber membranes by electrospinning technology. We realized the transformation of 2D membranes to 3D membranes by gas foaming technology. The gas is foamed by a controlled hydrolysis reaction of sodium borohydride, and the chemical reaction is: $NaBH_4 + 2H_2O \rightarrow Na_2O_2 + 4H_2\uparrow$. The expanded nanofiber scaffold was solidified and transferred to a vacuum freeze dryer.

**POSTER PRESENTATION**

# Impact of antifouling PEG Layer on the performance of functional peptides in regulating cell behaviors

Qi Chen, Runhui Liu
East China University of Science and Technology

The ultimate goal for regenerative medicine is to develop replacements for diseased tissues as alternatives to organ transplantation. Ideally, the biomaterials should be able to instruct cell behaviors, including adhesion, migration, proliferation, and differentiation, by modifying bioactive molecules (e.g., cell adhesion peptides (CAPs)) on material surfaces. However, only a handful of CAPs are identified and their functions are not fully understood at present. To identify new CAPs and explore their genuine interactions with specific cells, it is critical to decouple the nonspecific cell adhesion due to biofouling that is unfortunately universal unless antifouling strategies are applied. Polyethylene glycol (PEG) has been extensively used to provide the antifouling layer for covalent immobilization of CAPs. Nevertheless, the impact of PEG antifouling layer on CAPs' function is very less studies. The coexistence of CAPs-cell interaction studies using either no antifouling layer at all or PEG at variable chain length causes heterogeneous results that are difficult and even impossible to compare with each other.
Therefore, to address this overlooked long-lasting and critical problem, we provide a general guideline in selecting optimal PEG molecules (OEG8) for the identification, study, and application of CAPs and other functional peptides, which helps to avoid confusing and even controversial results, and fully display the genuine function of CAPs and other functional peptides.

References

Chen, Q.; Yu, S.; Zhang, D.; Zhang, W.; Zhang, H.; Zou, J.; Mao, Z.; Yuan, Y; Gao, C.; Liu, R.*, Impact of Antifouling PEG Layer on the Performance of Functional Peptides in Regulating Cell Behaviors. J. Am. Chem. Soc. DOI: 10.1021/jacs.9b07105

**POSTER PRESENTATION**

# Biodegradable hollow manganese/cobalt oxide nanoparticles for tumor theranostics

Qilong Ren
Donghua university

This article describes the fabrication of hollow manganese/cobalt oxide nanoparticles (MCO NPs) with tunable size through redox reaction and Kirkendall effect for cancer imaging and drug delivery. The MCO-70 NPs (with an average size of 70 nm) can act as glutathione (GSH)-responsive contrast agents for dual T1/T2-weighted magnetic resonance imaging (MRI). The degradation of MCO NPs by GSH led to the enhancement of their T1 and T2 signal by 2.24- and 3.43-fold compared with that of MCO NPs before degradation, respectively. Antitumor agents such as doxorubicin (Dox) could be encapsulated inside the cavity of the hollow MCO NPs (MCO-70-Dox) and be released in the presence of GSH. The MCO-70-Dox NPs showed good tumor growth inhibition effect in vitro and in vivo, which can be promising drug delivery vehicles and MRI contrast agents for tumor diagnosis and reporting drug release.

POSTER PRESENTATION

# "Transformed" $Fe_3S_4$ tetragonal nanosheets: a high-efficiency and body-clearable agent for magnetic resonance imaging guided photothermal and chemodynamic synergistic therapy

Guoqiang Guan
Donghua University

To retain the agents in tumors for cancer diagnosis and therapy, and then to remove them from the body, are key for the clinical applications of ideal inorganic theranostic agents. To meet these needs, we have developed a transformed theranostic platform, employing PVP coated $Fe_3S_4$ tetragonal nanosheets (TNSs), which could effectively accumulate in the tumor under magnetic targeting, whilst gradually transforming to small particles (~ 5 nm) over three weeks. These were then effectively excreted from the body in normal physiological conditions after exerting their therapeutic effect. The aqueous dispersion of PVP coated $Fe_3S_4$ TNSs had an intense near-infrared absorption, excellent photothermal conversion efficiency (64.3 %) and great T2 weighted magnetic resonance imaging properties (71.3 $mM^{-1}$ $S^{-1}$). In addition, $Fe_3S_4$ TNSs could realize a synergistic photothermal therapy (PTT)/chemodynamic therapy (CDT), because the localized heat produced by PTT from the defect-rich structure could enhance the Fenton process by utilizing the overproduced $H_2O_2$ in the tumor microenvironment, and in return, the produced •OH could inhibit tumor growth and recurrence after PPT. We thus developed a high-efficiency inorganic theranostic platform which was effectively cleared from the body. This will open up a new avenue for the design of inorganic agents for clinical applications in the future.

**POSTER PRESENTATION**

# Comparison of hydrophilic PEDOTs subjected to electrostimulation: toward a long-term implantable electrode

Yaqiong Zhang1, Qichao Pan1, Sihao Qian1, Ao Zhuang2, Bo Zhu1, Gao Qiu2
1.Shanghai University
2.Donghua University

Implantable microelectrodes for chronic neural stimulation require highly stable and biocompatible electrode materials. Conductive poly(3, 4-ethylenedioxythiophene)s (PEDOTs) have demonstrated a much lower impedance than Au due to their mixed ionic/electronic conductivity throughout the whole films. However, their application as implanting electrodes have been limited by their nonspecific interaction to proteins and cells, which are widely accepted as one of most critical reasons to the implanting induced foreign body reaction. To depress the nonspecific interaction of PEDOTs, the hydrophilic groups, i.e., phosphorylcholine, sulfobetaine, and oligo-ethylene glycol were grafted onto PEDOTs. All the previous works demonstrates that these hydrophilic PEDOT have an excellent performance to resist the nonspecific interaction of proteins and cells, indicating their potential used as electronic materials of bioelectrodes. However, to ensure the long-term implanting of electrodes, a stable protein/cell resistance is a must for their application in the implantable devices. Herewith, we carried out a systematic evaluation and comparison over the effects of electrical simulation on the protein/cell resistance of phosphorylcholine, sulfobetaine and oligo-ethylene glycol functionalized PEDOTs. As to ensure a stable low impedance is another key to promise the function of implanted bioelectrodes, we further compared the impedance stability of these hydrophilic PEDOT electrodes exposed to the electrochemical stress occurred along with the electrostimulation. All the experimental results indicated that the phosphorylcholine functionalized PEDOTs presents a superior stability at protein/cell resistance and interfacial impedance when subjected to electrostimulation.

**POSTER PRESENTATION**

# Ferulic acid-loaded nanofibers fabricated by coaxial electrospinning

Yaoyao Yang, Mian Wu, Kun Zhao, Ruiqing Wu
University of Shanghai for Science & Technology

There are many research hotspots in the field of biomedicine, such as, to improve the solubility of insoluble drugs, to regulate their release characteristics, and to improve their bioavailability and then, improve the efficacy. The ferulic acid (FA), an active component of traditional Chinese medicine, was dispersed in polymer fibers by electrospinning technology in this work. The effects of the sheath components on the drug release properties of nanofiber membranes prepared by coaxial electrospinning were studied.

A series of FA-loaded nanofibers, such as FA loaded PVP, FA loaded PVP-PVP and FA loaded PCL-PVP, were prepared by uniaxial and coaxial electrospinning. OM and SEM results show that the composite membranes are consist of nanofibers with fine diameter and smooth surface. X-ray diffraction (XRD) and infrared spectroscopy (FTIR) results show that FA in these nanofibers are amorphous instead of crystalline state, and the compatibility between FA and other components is good. In addition, the empirical curve between the FA concentration and the absorbance was established by ultraviolet spectroscopy (UV). It was used to calculate the release rate of FA in the vitro dissolution experiments of the FA-loaded nanofibers. The results showed that all FA-loaded nanofiber membranes could rapidly release the drug. FA loaded PVP nanofiber membranes and FA loaded PVP-PVP nanofiber membranes could complete the drug release in 5 minutes, while FA loaded PCL-PVP nanofiber membranes could complete 70% of the drug release in 5 minutes. At the same time, the release rate of FA from Chinses Angelica was analyzed and compared with these nanofibers. The results showed that the FA-loaded nanofiber membranes had much faster release rate.

In general, a series of nano-fiber membranes with fast drug release properties were prepared using FA as model drug. It has been improved that the obtained FA-loaded nanofiber membranes can promote the rapid dissolution of the insoluble FA. The work is valuable for the researches of fast drug delivery system of insoluble drugs.

POSTER PRESENTATION

# 131I 标记的聚磷腈纳米球的制备及其肿瘤诊疗应用

朱蔚 1, 赵凌舟 2, 范钰 1
1.东华大学
2.上海市第一人民医院核医学科

聚磷腈材料由于其独特的有机和无机杂化结构，不仅拥有优异的结构稳定性、耐热性、可化学修饰性，同时还具有优良的生物相容性，在生物医学领域受到了广泛的关注。本研究以六氯环三磷腈（HCCP）为交联剂，运用亲核取代反应与高分子材料支链聚乙烯亚胺（PEI）交联形成聚磷腈纳米球，并对其表面活性氨基进行功能化修饰，从而构建诊疗一体化纳米平台（如图 1 所示）。本工作首先通过沉淀聚合法将 HCCP 与 PEI 高度交联，形成表面富含氨基的纳米球，其平均粒径为 176 ± 25.6 nm，具有良好的分散性和稳定性。之后在纳米球表面通过化学键合 3-（4-羟基苯基）丙酸 N-羟基琥珀酰亚胺酯（HPAO），再修饰 1，3-丙烷磺酸内脂（1，3-PS），并对纳米球表面剩余氨基进行乙酰化处理。通过两性离子和乙酰基的修饰，提高了材料的抗蛋白吸附性能和生物相容性。通过 HPAO 表面的酚羟基螯合放射性核素 $^{131}$I，使纳米球可同时释放 γ 射线和 β 射线，从而实现对肿瘤的 SPECT 成像和放疗。在此基础上，建立乳腺癌荷瘤鼠模型，通过瘤内注射放射性核素 $^{131}$I 标记的聚磷腈功能化纳米球，发现材料在肿瘤部位可以实现较长时间的 SPECT 成像。通过抗肿瘤实验证明，放射性核素 $^{131}$I 标记的功能化聚磷腈纳米球与 Na$^{131}$I 及高分子材料 $^{131}$I-PEI.NHAc-HPAO-PS 相比具有更显著的肿瘤抑制效果，并能有效延长小鼠的生存期。上述结果证明，所制备的放射性元素标记的聚磷腈功能化纳米球具有良好的 SPECT 成像/放疗一体化性能，可潜在应用于肿瘤的精确成像诊断和治疗。

POSTER PRESENTATION

# 促创面愈合的新型脂肪干细胞多孔甲壳素纳米纤维支架的制备

刘颖
1. 东北师范大学
2. 北部战区总医院

目的：随着再生医学的发展，干细胞为创面损伤的治疗提供了一种可能，但没有良好的移植微环境将不利于细胞存活和发挥功能，且干细胞目前多为局部注射，存在给药方式受限而限制其使用。因此，本研究拟制备一种能够为创面应用干细胞的黏附、生长、繁殖和新陈代谢提供支持的生物活性材料。
方法：以鱿鱼软骨为原料，通过微纤维化技术并调控制备工艺制备系列尺度的甲壳素纳米纤维。通过表面改性技术调节甲壳素纳米纤维的表面性能，甲壳素纳米纤维直径在 5-10 纳米范围，脱乙酰度 10%-30%。以医用级 PLGA 微粒为模版，以甲壳素纳米纤维和胶原纤维复合制备新型多孔甲壳素纳米纤维支架材料；通过在支架上进行脂肪干细胞培养，细胞染色和细胞 GFP 标记后冰冻切片观察新型多孔甲壳素纳米纤维支架上细胞存活情况及对细胞形态的影响；通过小鼠建立大鼠创面模型，采用负载干细胞新型多孔甲壳素纳米纤维支架处理伤口；连续观察 10 天内创面面积、创面深度、愈合时间；HE 染色观察血管的数量和表皮迁移的程度、Masson 染色观察胶原生成情况；免疫荧光检测血管内皮生长因子 VEGF 表达情况；以评价探讨负载干细胞新型多孔甲壳素纳米纤维支架对伤口愈合的作用及机制。
结果：结晶紫染色结果表明，脂肪干细胞在支架上呈球状生长；gfp 标记后冰冻切片观察可见绿色荧光；与模型组相比，新型脂肪干细胞多孔甲壳素纳米纤维敷料能够在 7-10 天显著缩小创面面积（$p<0.05$），创面呈湿性愈合状态；HE 染色可见再生血管及表皮生长情况较模型组相比显著改善；Masson 染色观察胶原生成情况较模型组相比显著提高（$p<0.05$）；免疫荧光检测血管内皮生长因子 VEGF 表达增多。
结论：以甲壳素纳米纤维和胶原纤维复合制备的新型多孔甲壳素纳米纤维支架材料能够支持脂肪干细胞存活，其与干细胞形成的脂肪干细胞多孔甲壳素纳米纤维敷料能够促进创面愈合，研究结果将为创面治疗及再生医学提供新的基础及思路。

**KEYNOTE SPEECH**

# Electrospinning: a material fabrication technique for addressing Africa's current and future challenge

Samuel Chigome
Botswana Institute for Technology Research and Innovation

Electrospinning has the unique ability to produce continuous nanofibers from a wide range of materials in various fibrous assemblies. Due to their ability to form a highly porous mesh, their large surface-to-volume ratio and the ease of their fabrication, electrospun nanofibers have found applications in the development of products for a wide range of sectors. The relatively high production rate and simplicity of the electrospinning setup makes electrospinning highly attractive to both academia and industry. Therefore, electrospinning is seen as a material fabrication technique that can contribute significantly to addressing the major challenges that Africa is facing.

The presentation will give an overview of the progress made in the use of electrospinning to address Africa's challenges in the areas of health, energy and water. This will include the presenter's experience of electrospinning in Africa over the past 11 years, the central role that the Botswana Institute for Technology Research and Innovation (BITRI) and the African Materials Research Society (AMRS) are playing to promote the use of electrospinning in Africa and highlights of key electrospinning activities in the five regions of Africa.

**KEYNOTE SPEECH**

# Silk fibers and functional materials

Yaopeng Zhang
Donghua University

Animal silks produced by insects and spiders with hierarchical structures exhibit high strength and toughness and are of great interests for various medical applications. By mimicking the spinning process, spinning apparatus of silkworms and structure of silkworm silk, artificial silks tougher than natural silkworm silk were dry-spun from regenerated silk fibroin (RSF) aqueous solutions in air. Recombinant spider silk protein aqueous solution was also adopted to spin tough artificial spider silk via a bio-inspired microfluidic chip. Moreover, nanofillers such as carbon nanotubes, $TiO_2$ nanoparticles, graphene oxide nanosheets and carbon nanodots were applied to further reinforce the artificial silks and understand the structural evolution of the hybrid fibers. A green, sustainable, and promising route was developed to produce reinforced, ultraviolet resistant or fluorescent silk by feeding silkworm with the diet containing nanoparticles. We proposed a hierarchical structure model for silk fibers, in which the basic building block in the smallest nanofibril of silkworm was found to be of a ribbon shape with the thickness of about 0.4 nm. Two dimensional and three dimensional silk scaffolds with different structures, improved mechanical properties and multi-functions were prepared and applied for urethral reconstruction, nerve regeneration and liver function recovery.

**KEYNOTE SPEECH**

# The sustainability challenge for the fibre and textile industries

Xungai Wang
Deakin University

Sustainability has become a hot topic worldwide. Debates also rage on as to whether natural fibres or synthetic fibres are more sustainable, and which fibre or textile process is more environment friendly. Fibre blends have become prevalent, yet some common blends are notoriously difficult to separate and recycle. In this presentation, I will provide some insight into the current trend in this important area. I will also talk about the recent initiatives and research examples. I hope this presentation will stimulate some deep thinking among fibre researchers and students when exploring significant research questions and projects.

INVITED LECTURE

# Super strong and intrinsically fluorescent silkworms silk from carbon nanodots feeding

Suna Fan, Qi Zhan, Huili Shao, Yaopeng Zhang
Donghua University

Fluorescent silk fibroin (SF) fibers have great potential in biomedical application and special functions for marking and tracking. How to fabricate fluorescent SF fibers with good fluorescence stability by a simple and environmentally-friendly method has yet to be explored. Here we successfully produced fluorescent SF fibers by using silkworms as bioreactors to introduce rare-earth upconverting phosphors (UCPs) into silk fibroin. The modified silk exhibited bright green colors under 980 nm laser. Additionally, intrinsically fluorescent silk with reinforced mechanical properties were also fabricated via feeding silkworms with carbon nanodots (CNDs). The addition of CNDs hindered the conformation transformation, confined crystallization, and induced orientation of mesophase, resulting in reinforced silk fibers with breaking strength of 521.9 ± 82.7 MPa and breaking elongation of 19.2 ± 4.3%, improvements of 55.1% and 53.6%, respectively, in comparison with regular silk. The CNDs reinforced silk displayed intrinsic blue fluorescence when exposed to 405 nm laser and exhibited no cytotoxic effect on cells. This directly feeding method to produce fluorescent and reinforced SF fibers is green and environmentally friendly, and easy to use for mass production. Moreover, it provides an idea that SF fibers can be cooperated with more other materials, to improve the functionalities of conventional biomaterials and expand applications.

**ORAL PRESENTATION**

# Effect of alkali treatment on interfacial bonding in abaca fiber-reinforced composites

Ming Cai[1, 2, 3], Hitoshi Takagi[3], Li Yan[2]
1.Shanghai University of Engineering Science
2.Tongji University
3.Tokushima University

Abaca fibers demonstrate enormous potential as reinforcing agents in composite materials. In this study, abaca fibers were immersed in 5, 10 or 15 wt.% NaOH solutions for 2 h, and the effects of the alkali treatments on the mechanical characteristics and interfacial adhesion of the fibers in a model abaca fiber/epoxy composite system systematically evaluated. After 5 wt.% NaOH treatment, abaca fibers showed increased crystallinity, tensile strength and Young's modulus compared to untreated fibers, and also improved interfacial shear strength with an epoxy. Stronger alkali treatments negatively impacted fiber stiffness and suitability for composite applications. Results suggest that mild alkali treatments (e.g. 5 wt.% NaOH for 2 h) are highly beneficial for the manufacture of abaca fiber-reinforced polymer composites.

**KEYNOTE SPEECH**

# Engineering new materials from biomass. multiple applications of fibers in the context of a real sustainable approach

Jean Jacques Gaumet
University of Lorraine

We are in the center of an important paradigm shift concerning the limited availability of conventional resources such as fossil fuels and mineral raw materials (critical or non-critical), coupled with exponential population growth, all of which is leading to the potential risk of a worldwide energy and environmental crisis with societal ramifications.

Superior materials should be produced in a sustainable way, independent of fossil resources and affordable to all members of society. What has really encouraged research in biomass-derived carbon materials is the development of novel and sustainable products based on molecular structures available in nature.

Biomass originating from the photosynthesis of $CO_2$ and water and combined with solar light is the most efficient renewable carbon source on Earth, allowing for efficient production of biomass-derived carbon materials and nanomaterials [1].

Moreover, biomass-derived carbon materials are also being developed extensively in a variety of industries at both the micro and macro level, bringing forth solutions to the critical challenges we face in the near future. This should be always seen as a sustainable approach that can be promote within local, circular economies and local jobs. Note that the word “sustainability” has no meaning by itself and consequently it is fundamental to give “material sustainability” a context, depending on the environmental, economic and societal situation of an area [2]. This includes developing energy (production, conversion and storage), construction materials, paper and packaging, food, medical applications, transportation (e-mobility), and environmental applications (adsorbents, photocatalysis) [3-5].

Fibers can be seen as naturally derived from biomass or elaborated via various chemical or physical processes and some have been with our societies for thousands of years. They can be used directly or as part of composites with carbonaceous or inorganic matrices. Both environmental policy and environmental constraints stimulate innovation in using natural materials allowing developments of biopolymers and bio-composites. Thanks to the development of 3D printers, fibrous materials can be used at different scale from micrometric dimensions (e.g. medical applications) until the size of buildings. The mechanical properties are then often improved and this is a crucial challenge in aerospace and automotive structures.

We will present herein an overview about sustainable applications using various types of biomass waste for producing materials which range from nanosize to bulk structure. This will enable us to better estimate: 1) a product’s life cycle, 2) a product’s compatibility towards a desired solution, and 3) a product’s end of life scenario.

KEYNOTE SPEECH

# Artificial spider silk: a supertough fiber with a simple strategy

Jinlian Hu, Lin Gu, Yuanzhang Jiang
Institute of Textiles and Clothing, The Hong Kong Polytechnic University

Spider silks are tougher than almost all other materials in the world and thus are considered ideal materials by scientists and the industry. Although there have been tremendous attempts to prepare fibers from genetically engineered spider-silk proteins, it is still a very large challenge to artificially produce materials with a very high fracture energy because of the extremely low productivity and high cost. In this talk, a general introduction to spider silk, their key characteristics and their structures vs their properties will be introduced. A facile spider-silk-mimicking strategy using the chemical synthesis route will be reported for aciniform silk which is used by a spider for wrapping the prey and protecting their offsprings. A supertough (~387 MJ m-3) fiber, more than twice the reported value of a common spider dragline is obtained from this strategy. The toughness of this fiber is comparable to the value of the toughest spider silk, the aciniform silk of Argiope trifasciata. The existence of β-sheet crystals and α-helical peptides simultaneously in a pseudoprotein polymer is the core for such success.

**KEYNOTE SPEECH**

# Chemical control on glycopolymer self-assembly with immunological applications

Guosong Chen
Fudan University

Carbohydrates are the most abundant organic species in the world and also one of most important biological macromolecules with nucleic acids and proteins. The self-assembly of DNA and proteins make a significant contribution to our lives and they have been employed to make functional self-assembled materials. Compared to the development of DNA and proteins, our knowledge and manipulation to the self-assembly of carbohydrates as well as their functionality are quite limited. The major obstacle is the complicated chemical structure of oligosaccharides, i.e. perplexing glycoforms and micro heterogeneity on proteins, which make the research a problematic and long-term task. Under this circumstance, macromolecular self-assembly might provide an alternate insight to this problem. In this talk, I will present: 1) development of precise protein array with regular shape at mm scale controlled by protein-carbohydrate interaction; 2) construction of polymeric vesicles mimicking glycocalyx, structure, self-assembly and immunological functions; 3) control of macromolecular self-assembly by chemical reactions related to sugars.

INVITED LECTURE

# A strategy to construct self-reinforced nonswelling high-strength bacterial cellulose Hydrogels

Shiyan Chen, Minghao Zhang, Huaping Wang
Donghua university

A serious decline in mechanical properties of polysaccharide hydrogels caused by swelling has always been a difficult problem which greatly limited their application especially in the medical field. Herein, nonswelling high-strength natural hydrogels based on self-reinforced double-crosslinked bacterial cellulose (SDBC) were prepared. Inspiring by the concept of homogeneous composite materials, by regulating the ratio of LiOH/urea alkaline solution, the aggregation structure and nanostructure of SDBC hydrogels can be controlled, thereby constructed a uniquely nanofiber-network-self-reinforced (FNSR) structure and a new self-reinforcing mechanism is proposed. The prepared SDBC hydrogels have excellent mechanical properties at high water content (>91%) for the combination of double-crosslinking and uniquely FNSR structure, which can effectively prevent crack propagation and dissipate a large amount of energy. Especially, the compressive modulus can reach 3.17 MPa which is 56 times as that of native bacterial cellulose (BC). It is worthy to mention that no swelling occurs for the hydrogel, and the mechanical strength still remains excess 90% for 15 days in water, which is favorable for promising application in underwater equipment, implantable ionic devices, and tissue engineering scaffolds. This study also opens up a new horizon for the preparation of self-reinforced hydrogels.

ORAL PRESENTATION

# Phenolated wheat straw lignin based activated carbon fibers

Jian Lin
Beijing Forestry University

In China, most wheat straw were burnt to get energy, but the burning sometimes causes air pollution. We consider that wheat straw is an alternative biomass to wood in china. Thereby, utilization of wheat straw lignin (WSL) was investigated as one of wheat straw utilizations. WSL was obtained by autohydrolysis followed by enzymatic hydrolysis of wheat straw, and it was phenolated under acidic conditions. The highest yield of phenolated lignin, which had 32 % phenol and 3162 g/mol of molar mass determined by size-exclusion chromatography, were achieved by the reaction with WSL/phenol (L/P) ratio of 1/3 and 10 % $H_2SO_4$ at 120 °C for 2 hours. This reaction conditions were so severe as compared to the phenolation conditions for pine kraft lignin and sweetgum autohydrolysis/enzyme-treated lignin: much phenol and $H_2SO_4$ and higher reaction temperature were required. 31 P -NMR and HSQC exhibited that lignin substructures of β-O-4', β-5'/α-O-4', and β-β' were cleaved by the phenolation reaction. Concomitantly, phenol was incorporated at Cα position of lignin in addition to aromatic nuclei. As a result, aliphatic hydroxyl group was decreased, while the phenolic hydroxyl groups was increased by the reaction.

The phenolated WSL was attempted to be converted into carbon fibers via the following processes, melt spinning, thermostabilization and carbonization. The as-melt spun fibers could not be thermostabilized with air. The thermal stability of WAL fibers was improved by immersing the fibers in glutaraldehyde (GA) alkaline solution, which acted as a cross-linker for lignin molecules. Consequently, the GA-treated fibers was easily converted into infusible fibers by heating them in air to 250 oC at the heating rate of 5 oC /min. Finally, WSL-based carbon fibers (CFs) with the average diameter of 42 μm were obtained after the carbonization of the infusible fibers. The cross section of CFs were smooth, while the surface was rough, as shown in Fig. 1. The tensile strength of resultant CFs was around 490 MPa. The CFs were further converted into activated CFs (ACFs) with the BET specific surface area of 2010 m2/g by steam activation, indicating that WSL is a promising feedstock for precursor of ACFs.

KEYNOTE SPEECH

# Sustainable water purification using biomass nanofibers

Benjamin S. Hsiao
Stony Brook University

Nanoscale cellulose fibrous materials obtained from the chemical treatment of biomass are very effective agents for the removal of toxic species from water, including heavy metal ions. Our team at Stony Brook University has developed a simple, inexpensive and environmentally friendly approach to preparing nanostructured cellulose fibers for water purification, based on a nitro-oxidation reaction carried out on biomasses of diverse origins. There are three key advantages of the nitro-oxidation method. First, the method greatly reduces the consumption of chemicals, energy and water. Second, the processing effluent can be efficaciously neutralised to produce plant fertilisers. Third, the method is effective to extract nanostructured cellulose from underutilised raw biomass such as agriculture waste. The resulting nanocellulose is proven to be an efficient water purification material (membrane or adsorbent) that can treat a wide range of water pollution problems. The demonstrated technology represents an innovative means to enhance the nexus of food, energy, and water systems, and has many far-reaching impacts to improve quality of life.

**KEYNOTE SPEECH**

# Challenges for fabrication of fibers from nanocellulose

Daniel Söderberg
KTH Royal Institute of Technology

With the strong societal need for bio-based and biodegradable alternatives to today's plastic materials, there is a broad interest in fabricating fibers from cellulose. Cellulose-based fibers are clearly not a new concept, since e.g. Rayon, and more specifically viscose, has been around for a long time. However, these concepts are based on dopes from dissolved cellulose, and there is also a wide range of further developments of processing routes and technologies for fabricating regenerated cellulose fibers aiming at less harmful chemicals and improved properties.

As an appealing alternative to utilizing cellulose as in its basic polymeric form, there have been significant developments during the last years aiming at fabricating continuous fibers from nanocellulose, the structural nanocomponents of e.g. plants and wood, either as cellulose nanocrystals (CNC) or cellulose nanofibers (CNF). By utilizing the structural nanocomponent, the purpose is to fabricate fibers having a different performance envelope compared to fibers from regenerated cellulose, and maybe also as a competitor to high-performance synthetic fibers.

Recent results1-4 show that it is indeed possible to fabricate fibers from nanocellulose, from CNF as well as from CNC, and that the fabricated fibers can have unique capabilities allowing advanced material concepts enabling functionalization towards energy applications as well as for biomedical purposes. With respect to fabricating gibers from nanocellulose, the present challenge resides within the need for scaling-up manufacturing, which needs special considerations compared to the conventional spinning of fibers from regenerated cellulose, specifically the spinning of nanocellulosic high-performance fibers needs to be performed at very low concentrations5-6.

**KEYNOTE SPEECH**

# Pure Keratin Fibers from Poultry Feathers

Yiqi Yang
University of Nebraska-Lincoln， USA

Chicken and duck feathers were turned into pure keratin fibers via continuous wet spinning. Dissolution method was developed to turn the highly crosslinked keratin into linear structures with good spinnability in aqueous solution. Effective technologies were established to rebuild the crosslinkages in keratin after spinning to provide the regenerated protein fibers with good mechanical properties and aqueous resistance. The technology reported is suitable for all naturally crosslinked proteins, such as those from sorghum, camelina, milk, wheat and soybean.

INVITED LECTURE

# Animal silks as inspirations for smart fiber designs

Shengjie Ling
School of Physical Science and Technology, ShanghaiTech University

The brittle fibers are of great concern, as they often caused a sudden accidental rupture. Ductile and damage-tolerant fibers (DDTFs), hence, are desired in the multiple fields, such as aerospace engineering, civil engineering, and biomedical engineering. However, in practice, design and fabrication of DDTFs remain a significant challenge due to finite fiber size and limited processing techniques. In this regard, natural ductile and fracture resistance fibers can provide inspirations. Particularly, animal silks, hierarchically structured protein fibers, can teach us how to design DDTFs. In nature, animal silks feature an elegant tradeoff of mechanical strength, extensibility and damage tolerance, making them one of the toughest materials known. In this talk, we first introduce the strategies that spiders and silkworms design extraordinary fibers and then present our recent bioinspired and biomimetic approaches for developing mechanical and functional enhanced functional fibers.

ORAL PRESENTATION

# Critical areas of adhesion and proliferation of single cells revealed via material techniques of surface patterning

Xiang Yao, Jiandong Ding
State Key Laboratory of Molecular Engineering of Polymers, Department of Macromolecular Science, Fudan University, Shanghai 200438, China. E-mail: jdding1@fudan.edu.cn

Material cues to influence cell proliferation are a fundamental issue in the fields of Biomaterials, Cell Biology, Tissue Engineering and Regenerative Medicine. It is so far not clear whether or not a critical spreading area exists about cell proliferation, and how about the cell type dependence of spreading areas effect on cell proliferation. This literature aims to investigate proliferation of single mammal cells on micropatterned material surfaces. To this end, we prepared cell-adhesive circular microislands with 20 areas (20-8000 μm2) on the non-fouling background, and systematically examined adhesion and proliferation behaviors of different kinds of single cells (primary stem and non-stem cells, cancer and normal cell lines) on micropatterns.

In order to persistently confine single cells spreading, we developed a micropatterning technique to prepare circular microislands of arginine-glycine-aspartate (RGD) on a non-fouling PEG background. Primary rat MSCs, human cervix epithelial carcinoma cells (Hela, cancer cell line), mouse embryo fibroblast cells (NIH3T3, normal cell line) etc. were chose for the study of cell ahesion, proliferation and corrsponding critical areas. We further summarized the relative size relationship between critical areas about cell proliferation and the characteristic areas about cell adhesion both on patterned and non-patterned surfaces.

While proliferation of single primary cells was affected by cell spreading, cell lines, irrespective of normal and cancer cells, did not exhibit significant cell spreading effects. Based on the analysis of experimental data, we found two critical areas about cell proliferation: (1) the critical spreading area of cells from almost no proliferation to confined proliferation, denoted as AP, (2) the critical spreading area of cells from confined proliferation to almost freely proliferation, denoted as AFP. Besides, we also foud their (AP and AFP) relative size relationship with the characteristic areas of cell adhesion (Ac1, the critical area from apoptosis to survival for single cells; Ac2, the critical area of a microisland from single-cell adhesion to multi-cell adhesion; Asuspend, the project area of cells just falling onto tissue culture plates) obey a similar trend in our tested primary cells, but quit different from the trend of cancer and normal cell lines.

INVITED LECTURE

# Smart silk fibers with core-sheath structure and functional fillers

Bin Fei
The Hong Kong Polytechnic University

As the 'queen of fibers', silk is expected to find wider applications with new functions. Silk fibroin (SF) based fibers were prepared by wet spinning with a custom-made coaxial spinneret and smart functional loadings. The obtained core-sheath fibers were characterized on morphology, SF secondary structure, mechanical property and optical/electric functions. Fibers fabricated from 17 wt% SF/polyurethane solution presented the most regular morphology with homogeneous and circular cross-section. Double-layered hollow structure was observed, together with β-sheet conformation of SF. The fibers demonstrated higher absorbency than the raw silk and fine incorporation of long-lasting glowing pigments, conductive Ag nanowires, and cooling agents. The promising properties of resultant fibers indicate their possible applications in water or thermal management, phototherapy, and wearable electronics.

POSTER PRESENTATION

# Single molecular layer of silk nanoribbon as potential basic building block of silk materials

Qianqian Niu1, Suna Fan1, Huili Shao1, Xuechao Hu1, Huihui Zhang1, Rongliang Wu1, Benjamin S. Hsiao2, Yaopeng Zhang1
1.Donghua University
2.Stony Brook University

Silk is a kind of natural macromolecule material with good mechanical properties, and good biocompatibility. As the basic building block of the hierarchical structure, silk nanofibrils (SNF) is the key unit for the formation of high performance silk based materials. Conventional methods to prepare SNFs have some limitations, such as nanofiber aggregation/inadequate dissociation, low yield, toxic solvent, imperfect building model of silk hierarchical structure etc. To fabricate stable SNFs suspension efficiently, NaOH/urea system was used to prepare SNFs, and they are termed silk nanoribbons (SNRs) in this study. Transmission electron microscope and atomic force microscope were applied to confirm the size of SNFs accurately. It was found that SNRs possessed a lamellar (or ribbon) shape, which possessed an average thickness about 0.4 nm. The average width of SNRs was about 27 nm. This SNRs may be used as basic building blocks of silk materials.

POSTER PRESENTATION

# Super-strong and fluorescent silkworm silk from aggregation-induced emission (AIE) hexaphenylsilole feeding

Qi Zhan1, 2, Suna Fan1, 2, Yaopeng Zhang1, 2
1.State Key Laboratory for Modification of Chemical Fibers and Polymer Materials
2.College of Materials Science and Engineering， Donghua University

Fluorescent silk fibroin materials have wild applications in the biomedical field and bioassay system, but the aggregation caused quenching effect of fluorescent substances hinders their development. In this study, we reported an eco-friendly and economical feeding method to incorporate hexaphenylsilole (HPS), an important aggregation-induced emission (AIE) molecule, into silk. Results demonstrated that HPS was effectively incorporated into silk fibroin instead of silk sericin. This method altered the second structures, reduced cross-sectional area, crystallinity and crystallite size of silkworm silk. In addition, the obtained silk exhibited strong mechanical properties with breaking elongation of 11.9 ± 1.7% and breaking strength of 503.0 ± 44.4 MPa improvement of 17.8% and 54.0%, respectively, in comparison with the regular silk. The functional silk may pave a route to prevent the quenching effect of conventional fluorescent silk fibroin materials.

**POSTER PRESENTATION**

# Top-down peeling bacterial cellulose to high strength ultrathin film and multifunctional fibers

Zhuotong Wu, Shiyan Chen, Rongliang Wu, Nan Sheng, Minghao Zhang, Peng Ji, Huaping Wang
Donghua University

In billions of years of evolution, nature produced complex high-performance structural materials, such as wood, spider silk, and bone. These materials exhibit excellent properties due to main reasons: the materials with highly-aligned tight arrangement of the building blocks. And, strength and toughness are regarded as mutually exclusive in materials.

Bacterial cellulose (BC) is a three-dimensional network hydrogel consisting of layers of disordered nanofibers assembled by bacteria. BC could give full play to the excellent properties of nanofibers. However, it is difficult to realize the nanofiber alignment for the strong hydrogen bonding between the nanofibers.

Here, we first adopted an easily realized method to obtain high strength ultrathin BC film with highly-aligned tight nanofiber structure. First, the hydrogen bond between nanofibers was weakened by simple solvent replacement, and then we obtained the high-aligned nanofiber BC films by stretching the hydrogels. After drying by hot pressing, we first realized the peeling of BC film layer by layer to obtain the high strength ultrathin films inspired by graphene peeling method. The obtained film can reach the tensile strength of 758 MPa and the toughness of 42.3 MJ m-3. By further twisting the films, very strong and tough fibers were obtained with a high strength of 954.2 MPa and the record high toughness of 93.2 MJ m-3. Furthermore, functional BC fibers can be achieved by embedding functional materials (dyes, functional nanomaterials, carbon nanotubes, etc.) in the ultrathin films before twisting. This paper first realized the transformation from BC gel to high strength and high toughness ultrathin BC film and fiber. it opened up a new platform for the application of BC gel as high-strength flexible electronic device substrates, medical fibers, and for wearable electronics.

**POSTER PRESENTATION**

# Influence of chemical modifications on tensile properties and microstructure of abaca fibers

Xian Zhang
Shanghai University of Engineering and Technology

The natural fibers play a major role in reinforcement composites due to their important properties such as lightweight, biodegradability and non-toxicity. Abaca fiber is one of such material which is cheap and available easy, with high cellulose. There have been some research works on the tensile properties of nature fibers with little attentions on the relationship between the variations of lumen and tensile properties. In this study, the abaca fibers were treated with 5% NaOH solution for 0min, 5min, 10min, 20min and 30min, and the effects of alkali treatments on the fiber properties were explored and the structural changes of lumen were also observed.

POSTER PRESENTATION

# RHMPAs based on PPC for reactive hot-melt polyurethane adhesives

Jiqing Huang1, Yongxi Cheng1, Zenghe Liu2, Zhengwei You2
1.Beijing Institute of Aerospace Testing Technology
2.State Key Laboratory for Modification of Chemical Fibers and Polymer Materials， College of Materials Science & Engineering， Donghua University

Reactive hot-melt polyurethane adhesives (RHMPA) is a compelling green adhesive without any solvent. However, current RHMPAs are mainly made from non-renewable petroleum-based polyols. Poly (propylene carbonate) (PPC) derived from abundant, cheap and renewable greenhouse gas $CO_2$ is a high-profile green renewable polyol. Here, a series of RHMPAs (PPC-RHMPAs) based on PPC with different carbonate linkage content were prepared. The difference of carbonate linkage content of PPC-RHMPAs was investigated by 1H-NMR and FTIR. The effects of carbonate linkage content on the moisture-curing rate, crystallization behavior, thermal stability, mechanical properties and adhesion properties of the resulted PPC-RHMPAs were analyzed and discussed by real-time FTIR, DSC, TGA, tensile and adhesion tests. This work would pave a way for designing high-performance PPC-RHMPAs.

POSTER PRESENTATION

# 生物基呋喃聚酯纳米纤维的制备及其应用

李潇然[1,2]，李振环[1,2]，苏坤梅[1,2]
1. 天津工业大学
2. 天津工业大学省部共建分离膜与膜过程国家重点实验室

生物质资源是人类可开发利用的理想型资源，利用六碳糖或纤维素等生物质资源制备的 2，5-呋喃二甲酸(FDCA)具有和对苯二甲酸(PTA)极为相似的结构;用生物质资源 FDCA 替代化石资源 PTA 制备的呋喃基聚酯新材料，可实现原料绿色、加工绿色、应用绿色和产品绿色，对实现人类生产的可持续发展有重要意义。分别以甲酯化的 2，5-呋喃二甲酸甲酯(DMFDCA)和 FDCA 为聚合单体，合成了聚(2，5-呋喃二甲酸-1，4-丁二醇酯) (PBF) 和聚(2，5-呋喃二甲酸-乙二醇酯)(PEF)。探讨了聚合单体对聚酯分子量的影响，通过特性粘度表征确定了聚合程度。以三氟乙酸(TFA)和二氯甲烷(DCM)为溶剂对 PBF 进行静电纺丝，利用 FT-IR、DSC、TG、UVC、SEM、及力学性能测试等多种表征手段，研究了 PBF 纳米纤维的性能和结构。结果发现：在纺丝电压 20kv、接收距离 15cm、纺丝溶液浓度 10wt％时可以得到适合于空气过滤的纤维。通过分析，PBF 纳米纤维直径在 400-500nm 之间，膜孔径分布在 2nm 左右，具有高孔隙率、耐酸不耐碱性、一定的紫外光耐受性以及优异的空气过滤性能。进一步对 PEF 和 PBF 的合成工艺中试放大，分析聚合过程中的温度变化曲线，发现合成的关键酯交换温度为 PEF 168℃、PBF 154℃，缩聚温度为 PEF 203℃、PBF 215℃。将 PBF、PEF、PBT 和 PET 共同纺织成纳米纤维发现其纤维膜均表现为疏水性；进一步测试空气过滤能力，发现 PBF、PEF 纤维膜的过滤效率为 99.17％、99.26％，性能略低于 PET、PBT。经过驻极改性增加静电吸附作用后，PEF、PBF 纳米纤维膜的过滤效率显著增加，达到了满意效果。
关铡词：2，5-呋喃二甲酸；呋喃基聚酯；静电纺丝；纳米纤维；空气过滤

参考文献：

李立博，苏坤梅，李振环，et al.呋喃基聚合物的研究进展[J].石油化工，2017，（8）：1080-1088
Diao L, Su K, Li Z, et al. Rsc Advances, 2016, 6(33):27632-27639.

**KEYNOTE SPEECH**

# Up-scalable nanofibrous membrane materials for diverse applications in liquid/water separation and purification

Dong Wang, Ke Liu, Pan Cheng, Qihao Guo, Mufang Li
Wuhan Textile University

Membrane-based separations for water purification, disinfection and desalination have been increasingly applied to address the global challenges of water scarcity, the pollution of aquatic environments and industrial separation (such as biopharmaceutical purification). However, progress in water-purification/liquid-separation membranes has been constrained by the inherent limitations of conventional membrane materials mainly involving inefficient permeability–selectivity trade-off and lack of functionality.

Various methods have been widely adopted to control the structure and chemical functionality in polymer films in order to contrive new classes of membranes for water purification. We developed a novel technique to large-scale prepare the polymeric nanofibers and their coating layer based membrane, which possesses some advantages against other membranes including: (1) fiber-stacking based process facilitating the formation and tunability of pore structure of membrane, (2) high specific surface area beneficial to the functionalization in the preparation or post-modification of membrane, (3) interconnected pore structure favoring the water penetration, (4) ultrathin active nanofiber layer (normally 5μm) with high uniformity of membrane pore size conducive to the separation of substances from water, (5) melt-spinning and aqueous-suspension coating makes the technique more eco-friendly and easy for industrialization.

Beneficial from above merits, nanofibrous membranes with diverse structures have been designed to extend the applicability of them in domestic water and industrial water filtration/separation fields. (1) Microfiltration membrane with average pore size varied in range of 0.05-1 μm present high filtration efficiency with excellent water permeability and unique anti-fouling property, showing a superiority in high-efficient interception of particulate contaminates or microbes in drink water and environmental water body, as well as the purification and sterilization in biopharmaceutical, microelectronic manufacture and food industries. (2) It has been employed as the support scaffold to immobilize the (metal-alloy or ceramic) nano-architecture for the catalytic degradation in environmental engineering or catalyzed synthesis in chemical engineering, and to introduce chemical groups on membrane surface for chemical adsorption or affinity separation in protein purification process. (3) It has been also utilized as the substrate for the preparation of selective layer of nanofiltration or/and RO membranes with a distinct improvement for the optimization of trade-off factor in NF and RO process. Therefore, the nanofibrous membranes have exhibited tremendous potential in various separation and purification applications.

KEYNOTE SPEECH

# Polymer fibril network morphology enables high-performance organic solar cells

Yanming Sun
Beihang University

The last decade has witnessed the rapid progress of bulk heterojunction (BHJ) organic solar cells (OSCs) and now high power conversion efficiencies (PCEs) exceeding 16% have been realized in different types of organic solar cells. However, the relationship betweer molecular structure-BHJ morphology-device performance remains unclear. Our group mainly focuses on the development of high-performance benzodithiophene-4, 8-dione (BDTDO)-based polymer donors through molecular backbone and side chain engeering. We found that BDTDO-based polymer donors easily form fibrils of nanometers with a long axis parallel to conjugated backbone. We systematically studied the relationship between polymer strcture, morphology and photovoltaic performance. Finally we showed that the formationof well-defined fibrillar structure is a promising approach to achieving a favorable BHJ morphology, which can yield high PCEs for OSCs.

**ORAL PRESENTATION**

# Novel living polymer/bacteria composites for bioremediation

Mahsa Mafi, Andreas Greiner
University of Bayreuth

Encapsulation of Micrococcus Luteus (M. Luteus) in polymeric composites has been employed for the bioremediation, sequestration of metals and for the biodegradation of chemical pollutants and toxic components in waste water. Polymer composites in the form of nonwovens of nanofibers, or core/shell particles can provide a bacterial friendly environment for transfer of nutrients and metabolisms, with the least leakage of bacteria. M. Luteus is encapsulated in a hydrophilic core of poly (vinyl alcohol), following by synthesis or coating of a proper shell as a support to maintain the chemical and mechanical strength. The biological activity of bacteria is confirmed by Live/Dead analysis and agar plate tests. SEM and TEM analysis were utilized for morphological studies of polymer composites. As a result of the successful encapsulation of the alive bacteria in polymers, longer storage time in their functional state were achieved.

**ORAL PRESENTATION**

# Novel living polymer/bacteria composites for bioremediation

Mahsa Mafi, Andreas Greiner
University of Bayreuth

Encapsulation of Micrococcus Luteus (M. Luteus) in polymeric composites has been employed for the bioremediation, sequestration of metals and for the biodegradation of chemical pollutants and toxic components in waste water. Polymer composites in the form of nonwovens of nanofibers, or core/shell particles can provide a bacterial friendly environment for transfer of nutrients and metabolisms, with the least leakage of bacteria. M. Luteus is encapsulated in a hydrophilic core of poly (vinyl alcohol), following by synthesis or coating of a proper shell as a support to maintain the chemical and mechanical strength. The biological activity of bacteria is confirmed by Live/Dead analysis and agar plate tests. SEM and TEM analysis were utilized for morphological studies of polymer composites. As a result of the successful encapsulation of the alive bacteria in polymers, longer storage time in their functional state were achieved.

**ORAL PRESENTATION**

# Fabrication of amidoxime-modified cellulose fiber membranes and their application in uranium extraction from seawater

Wangliang Li[1], Yanxiang Li[1], Ying Wang[1, 2]
1.Institute of process Engineering
2.University of Chinese Academy of Sciences

As a kind of nuclear material, uranium is a very important guarantee for the development of nuclear energy in the future. There are abundant uranium resources in the seawater, uranium extraction from seawater is regarded as one of the seven chemical separations to change the world [1].

Due to the high specific surface area, strong mechanical property and good processing property of fabric materials, modified polymeric fiber-based adsorbents have been recognized as the promising approach for the uranium extraction from seawater [2]. Compared with the traditional polymeric fibers, natural cellulose has the characteristics of abundant sources, low cost, non-toxicity, harmlessness, degradability, renewability and hydrophilicity. In this work, cellulose fibers were modified with amidoxime groups followed by vacuumed-filtering and hot-pressing into nonwoven membranes. On the one hand, the form of fibrous membranes avoided the loss of adsorption materials and achieved the continuous dynamic operation process of the adsorption. On the other hand, the amidoxime-modified cellulose fiber has adsorption with good capacity and selectivity on uranium. Both of the static and dynamic adsorption of uranium were fitted and analyzed. The static adsorption process satisfies the Langmuir model and Pseudo second-order kinetic model. The dynamic process conforms to the Dose-Response model. In addition, the cellulose fiber membranes can be eluted, regenerated and recycled.

ORAL PRESENTATION

# Construction of cadmium sulfide based semiconductor heterojunctions on carbon fibers for purifying wastewater under visible-light irradiation

Yan Zhang, Lisha Zhang, Zhigang Chen
State Key Laboratory for Modification of Chemical Fibers and Polymer Materials， Donghua University

Cadmium sulfide (CdS)-based semiconductor heterojunction powders have exhibited the enhanced photocatalytic activities, but their practical applications have been limited due to their poor recycling from flowing wastewater. To solve these problems, with carbon fiber (CF) as the fixing substrate, we constructed two CdS-based heterojunctions ($MoS_2$/CdS and $TiO_2$/CdS) as two models on CF surface for purifying wastewater under visible-light irradiation. Both of CF/$MoS_2$/CdS and CF/$TiO_2$/CdS display strong and wide photo-absorption. Under visible-light irradiation, CF/$MoS_2$/CdS cloth (area: 4×4 $cm^2$) can expeditiously remove 97.3% rhodamine B, 55.6% tetracycline hydrochloride and 68.4% hexavalent chromium after 100 min, and 97.3% methylene blue after 70 min, while CF/$TiO_2$/CdS cloth (area: 4×4 $cm^2$) can degrade 95.44% methylene blue, 64.95% acid orange 7, 91.37% tetracycline hydrochloride and remove 90.70% hexavalent chromium after 120 min. Therefore, both CF/$MoS_2$/CdS and CF/$TiO_2$/CdS have the capacity as recyclable photocatalysts with high efficiency for purifying wastewater.

**ORAL PRESENTATION**

# Flexible and washable photothermal nonwoven fabric for solar-driven seawater desalination

Bo Zhu, Zhigang Chen
State Key Laboratory for Modification of Chemical Fibers and Polymer Materials, College of Materials Science and Engineering, Donghua University

Solar-driven seawater desalination has been considered to be an attractive methods to alleviate the global freshwater scarcity [1]. Up to now, various types photothermal membranes were well developed, and they have made great achievements and breakthroughs in solar-driven seawater evaporation [2]. However, their pragmatic applications are often limited by substantial salt accumulation. To solve this issue, we have designed and prepared flexible and washable CNT-embedded polyacrylonitrile (PAN) nonwoven fabrics by a simple electrospinning route [3]. The wet fabric exhibits a strong photoabsorption in a wide spectral range (350-2500 nm), and it has a photoabsorption efficiency of 90.8%. When coated onto a PS foam, the fabric shows a high seawater evaporation rate of 1.44 kg m-2 h-1 under simulated sunlight irradiation (1.0 kW m-2). With a high concentration of simulated seawater as the model, the accumulation of solid salts can be clearly observed on the surface of the fabric, resulting in a severe decay of the evaporation rate. These salts can be effortlessly washed away from the fabric through a plain hand-washing process. The washing process has a negligible influence on the morphology/photoabsorption /evaporation performance of the fabric, demonstrating good durability. More importantly, a larger fabric can easily be fabricated, and the combination of washable fabrics with various parallel PS foams can facilitate the construction of large-scale outdoor evaporation devices, conferring the great capacity for efficient desalination of seawater under natural sunlight.

References:

[1] Elimelech, M.; Phillip, W. A., The Future of Seawater Desalination: Energy, Technology, and the Environment. Science, 2011, 333, 712-717.
[2] Gao, M.; Zhu, L.; Peh, C. K.; Ho, G. W., Solar Absorber Material and System Designs for Photothermal Water Vaporization Towards Clean Water and Energy Production. Energy Environ. Sci. 2019, 12, 841-864.
[3] Zhu, B.; Kou, H.; Liu, Z.; Wang, Z.; Macharia, D. K.; Zhu, M.; Wu, B.; Liu, X.; Chen, Z., Flexible and Washable CNT-Embedded PAN Nonwoven Fabrics for Solar-Enabled Evaporation and Desalination of Seawater. ACS Appl. Mater. Inter., 2019. (DOI: 10.1021/acsami.9b12806)

KEYNOTE SPEECH

# Assembly of fibers from colloidal liquid crystals based on low-dimensional materials: preparation and properties

Shenmin Zhu, Hui Pan, Jingyi Lin, Yao Li
State Key Lab of Metal Matrix Composites， School of Materials Science and Engineering， Shanghai Jiao Tong University

With the rapid development of low-dimensional materials, colloidal liquid crystals have recently attracted more and more attention. Due to the unique physical properties, novel low-dimensional colloidal liquid crystals (CLCs), such as carbon nanotubes and graphene, have shown wide potential applications in electronic devices, sensors, energy storage devices, etc. As the macroscopic assembly of CLCs is one of the most effective strategies to construct ordered 3D architectures, CLC exploration is of theoretical and practical significance. Although traditional inorganic mineral CLCs have been studied nearly for a hundred years, the investigation on novel low-dimensional CLCs is still in the primary stage. Our work focuses on the design, preparation and application of new CLCs based on low-dimensional materials. Through macroscopic assembly, flexible films and fibers were fabricated which were further used as anodes in lithium batteries. Due to the ordered structure inherited from their CLCs, the composite exhibited excellent electron transfer and ion diffusion characteristics which benefit the performance of the lithium batteries.

INVITED LECTURE

# Printed flexible and wearable energy storage devices

Yuanlong Shao
Soochow University

Recent advancements in portable electronics and emerging interest in wearable miniaturized devices, on-body medical monitor and Internet of Things (IoT) nodes have propelled the needs for miniaturized energy storage devices1. Innovations in electrode materials and development in scalable approaches to these new flexible and miniature energy storages are required before expanded and widespread implementation on body, curvilinear objectives and logistics packaging and transportation. Thus, these small-scale and discrete energy storage devices need to be reasonably thin, mechanically robust, electrochemically durable and dynamically flexible in nature2. However, feasible designs and viable manufacturing means of fabricating miniaturized energy storage devices, such as micro-supercapacitors (MSCs), for energy delivery or energy harvesters with high power capabilities remains a significant challenge3-4. An array of cost-effective manufacturing processes for MSCs seems poised for addressing the pressing needs. Notable innovations include laser scribing, electrochemical deposition, vacuum filtration, spray coating. These pioneering demonstrations, while promising, cannot precisely print out elaborated designs of micro-patterns, especially when more than one type of active materials need to deposit, a key prerequisite for high electrochemical performance of MSCs. Meanwhile, inkjet printing, characterized by the formation of material-containing droplets by a sudden pressure pulse in the nozzle chamber, offers a promising strategy to create complex layouts of multi-functional materials with well-defined features, periodicities and morphologies and scalability over large areas. Herein, we will present our recent works in the field of printed flexible and wearable energy storage devices, including the inkjet printed 1T $MoS_2$ based supercapacitor and VN based self-power energy storage unit.

References:

[1] Y. Shao, M. F. El-Kady, L. J. Wang, Q. Zhang, Y. Li, H. Wang, M. F. Mousavi and R. B. Kaner, Chem. Soc. Rev., 2015, 44, 3639-3665.
[2] M. F. El-Kady, Y. Shao and R. B. Kaner, Nature Rev. Mater., 2016, 1, 16033.
[3] Y. Shao, M. F. El-Kady, J. Sun, Y. Li, Q. Zhang, M. Zhu, H. Wang, B. Dunn and R. B. Kaner, Chem. Rev., 2018, 118, 9233-9280.
[4] Y. Shao, J. Li, Y. Li, H. Wang, Q. Zhang and R. B. Kaner, Mater. Horiz., 2017, 4, 1145-1150.
[5] Z. Tian, X. Tong, G. Sheng, Y. Shao, L. Yu, V. Tung, J. Sun, R. B. Kaner, Z. Liu, Nature Communications, 2019, in press.

**ORAL PRESENTATION**

# Hanging photothermal fabrics for solar seawater desalination

Zixiao Liu, Zhigang Chen
State Key Laboratory for Modification of Chemical Fibers and Polymer Materials, College of Materials Science and Engineering, Donghua University, Shanghai 201620, China

Solar-enabled evaporation for seawater desalination has been considered to be an attractive, renewable and environmentally friendly way to alleviate the long-standing problem of freshwater scarcity. [1] Till now, most of the studies focus on floating photothermal membranes on seawater and generate water vapor under solar irradiation [2], suffering from three problems including unavoidable heat loss, single-face evaporation (only top surface) and solid-salt separation. To solve these problems, we have fabricated PANi-cotton photothermal fabric and come up a hanging model that lifts the photothermal fabric away from seawater with only two fabric edges immersed in seawater. [3] In this model, the top and bottom surfaces of the fabric are surrounded by air. Thus, the fabric can evaporate seawater on its two surfaces, effectively overcoming the limitations associated with the floating model. Meanwhile, the coverage of air makes the heat dissipation is only 16.8% of that from floating fabric, due to the extremely low conductivity (0.023 W m-1 K-1) of air. More importantly, the hanging fabric is arc-shaped, with its two edges higher than the center. Seawater is absorbed by the fabric from two sides to the center. With the progress of solar-driven seawater evaporation, the high-salinity brine is gathered at the lowest center of hanging fabric and finally dropped down for collection. Dropping brine takes away the salt in fabric, and thus the solid salt separation can be completely avoided. Additionally, the brine is also meaningful for other industry applications, such as sea-salt preparation or the chlor-alkali industry.

[1] Elimelech, M.; Phillip, W. A., The Future of Seawater Desalination: Energy, Technology, and the Environment. Science., 2011, 333, 712-717.
[2] Tao, P.; Ni, G.; Song, C.; Shang, W.; Wu, J.; Zhu, J.; Chen, G.; Deng, T., Solar-driven interfacial evaporation. Nat. Energy., 2018, 3, 1031-1041.
[3] Liu, Z.; Wu, B.; Zhu, B.; Chen, Z.; Zhu, M.; Liu, X., Continuously Producing Watersteam and Concentrated Brine from Seawater by Hanging Photothermal Fabrics under Sunlight. Adv. Funct. Mater., 2019, 1905485.

**ORAL PRESENTATION**

# Janus fibrous membranes for water treatment by membrane distillation

Shuo Cong, Zeman Liu, Fei Guo
Dalian university of technology

In the present study, a novel Janus monocomponent nanofibrous membrane test for AGMD performance was fabricated by initiated Chemical vapor deposition (iCVD) technique and electrospinning using the polyacrylonitrile (PAN). The Janus nanofibrous membrane was composed of two layers: the iCVD treated superhydrophobic electrospun nanofiber layer as the top layer (i.e., the active layer facing the feed side) and as-spun hydrophilic electrospun nanofiber layer as the bottom (support) layer which faced the permeate side (i.e., the air gap in this study). The Janus fibrous membranes were tested in a lab-scale AGMD system. The effect of with and without the hydrophobic layer and thickness ratios of the two layers on the AGMD performances in terms of permeate flux, salt rejection ratio, and mass transfer coefficient were investigated systematically.

**KEYNOTE SPEECH**

# The rising fiber electronics

Huisheng Peng
Fudan University

It is critically important to develop miniature electronic devices for a variety of emerging new fields, such as wearable facilities, smart textiles and internet of things. Here a novel family of energy and electronic devices with various functionalities, e.g., energy harvesting, energy storing, lighting, and sensing, in 1D fiber configuration are carefully discussed with unique and promising advantages such as lightweight and weaveable compared with the conventional planar architecture. The main efforts will be made to highlight the recent advance in the electrode material, device structure and property extension.

**INVITED LECTURE**

# 静电纺聚合物纳米纤维锂电池隔膜

缪月娥, 雒香, 朱晓波, 刘天西
东华大学

隔膜材料作为电池的关键组件之一，在电池中起着隔离正、负极直接接触，阻止电池内的电子传输而允许电解液离子自由通过的作用。隔膜性能的优劣在根本上决定了电池的界面结构与内阻，并直接影响着电池的容量、循环性能以及安全性能等特性。静电纺聚合物纳米纤维具有原料种类可选范围广、直径可控、孔隙率和孔径分布可调等优点，其作为锂电池（如锂离子电池、锂硫电池等）隔膜时相较于传统聚烯烃多孔膜有望获得更高的吸液率、保液性和离子电导率，进而提高电池的比容量和循环性能等。因此，本工作采用静电纺丝技术分别制备了聚酰亚胺、含氟聚酰亚胺、羧基功能化的聚酰胺酸纳米纤维膜以及孔径结构可控的聚丙烯酸/聚丙烯腈复合纳米纤维膜，并将其用作锂电池隔膜。结果表明，聚酰亚胺纳米纤维隔膜相较于商用 Celgard 隔膜具有更好的热稳定性和电解液浸润性，由其所组装的锂离子电池展现了更高的倍率和循环稳定性。此外，羧基功能化和含氟官能团的引入使聚合物纳米纤维膜在用作锂硫电池隔膜时不仅促进了正电性锂离子的迁移速率，还有效构造了负电场、抑制了负电性多硫化锂的穿梭效应，从而显著提高了锂硫电池的循环稳定性和倍率性能，为高性能锂硫电池隔膜的设计和开发提供了新思路。

ORAL PRESENTATION

# A single integrated 3D-printing process customizes elastic and sustainable triboelectric nanogenerators for wearable electronics

Shuo Chen, Han Zuo, Zhengwei You
Donghua University

Triboelectric nanogenerator (TENG) devices have gotten great attention in wearable power sources and physiological monitoring. However, the complicated assembling and the molding processing retard their applications. Here, 3D-printed TENGs (3DP-TENGs) are designed and readily fabricated by a single integrated process without additional assembling steps. The TENGs contain poly(glycerol sebacate) (PGS) and carbon nanotubes (CNTs) as the two electrification components. Conductive CNTs also serve as electrodes. Elastic PGS matrix makes TENGs intrinsically responsive to biomechanical motions leading to robust energy outputs. The hierarchical porous structure of the 3DP-TENG results in higher output efficiency than traditional molded microporous TENG counterparts. TENGs with different 3D shapes are readily fabricated for different applications. The 3DP-TENG insole efficiently harvests biomechanical energy to drive electronics. A ring-shaped TENG acts as a self-powered sensor to monitor the motion of fingers. Furthermore, the use of bio-based and biodegradable PGS matrix combining with efficient recycle of CNTs makes 3DP-TENGs favorable from sustainable perspective. This work provides a new strategy to design and tailor 3D TENGs that will be very useful for diverse electronic applications.

Reference

[1] Chen, S., Huang, T., Zuo, H., et al. A single integrated 3D-printing process customizes elastic and sustainable triboelectric nanogenerators for wearable electronics. Adv. Funct. Mater., 2018, 28, 1805108.
[2] Sun, L., Chen, S., Guo, Y., et al. Ionogel-based, highly stretchable, transparent, durable triboelectric nanogenerators for energy harvesting and motion sensing over a wide temperature range. Nano Energy, 2019, 63, 103847.

**ORAL PRESENTATION**

# Wearable devices of nanofibers coated yarns (NCY)

Mike Tebyetekerwa1, Shengyuan Yang2, Meifang Zhu2
1.The Australian National University
2.College of Materials Science and Engineering， Donghua University

Traditional textile yarns are known to be super electrical insulators. However, for wearable smart textile applications, these yarns are required to be good electrical conductors. In wearable energy storage applications such as batteries and supercapacitors and sensor applications, the employed yarns are usually coated with conducting active materials using a variety of techniques such as in-situ solution/vapor phase polymerization, dip coating, thin film coating using layer by layer assembly, atomic layer depositions and electrochemical depositions. However, these methods are known to be hectic, uncontrollable and hardly scalable. Beyond these, they also give brittle coatings which tend to crack easily if coated yarns are incorporated in traditional textiles during use or even during post manufacturing in weaving/knitting and sewing processes.

In this talk, i will explain a new simple, scalable, quick single-step and controllable coating process which involves winding functional nanofibers onto textile spun yarns to give a fully flexible high surface area and all-textile tough energy storage and sensor yarns for wearable electronics. The technique is an improved method of functional yarn-coating employing modified traditional electrospinning. Brieflly, the nanofibers are continuously coated on the textile filament acting as a collector to form a uniform controlled layer via electrospinning. The resultant nanofibers coated yarns (NCY) when assembled into yarn-based devices showed excellent electrochemical performance and great durability. Yet still, when woven, knitted and braided into traditional textiles, no significant capacitance difference was shown. Moreover when NCY conducting yarns were employed as organic vapour sensors, they showed good response and durability functions.

The method i intend to present during the conference is highly promising to opening up the new strategy to fabricate all-textile, fully-flexible smart yarns which perfectly fit in the knitting, weaving and braiding mature textile industries for use in wearable smart textiles. And beyond energy storage and sensor applications, the technique can be employed to obtain all-textile energy harvesting yarns (NCY solar cells, nanogenerators, etc).

**ORAL PRESENTATION**

# Enhancing Pervaporation Dehydration Performance by Manipulating the Interfacial Interactions via Introducing Polyethyleneimine Modified Graphene Oxide Intermediate Layer

Peiyun Li, Xuefen Wang
Donghua University

A novel three-tier composite membrane based on highly porous nanofibrous substrate was demonstrated for efficient isopropanol dehydration by pervaporation. Polyethyleneimine (PEI) modified graphene oxide (GO) sheets were vacuum-assistant assembled onto porous electrospun polyacrylonitrile (PAN) nanofibrous substrate to achieve a smooth, hydrophilic and compact PEI-GO intermediate layer. The introduction of PEI chains endowed GO interlayer with sufficient interaction for bonding adjacent GO nanosheets to enhance stability in water/isopropanol mixture and also with the ascended interlamellar space to improve the water-sorption ability due to the abundant active amino groups. Benefiting from PEI-GO layer, a defect-free sodium alginate (SA) skin layer could be facilely manufactured with elaborately controlled thickness as thin as possible in order to reduce mass transfer resistant and enhance permeability maximally. And the interlayer contributed to enhance interfacial adhesion to promote the structure integrity of three-tier thin-film nanofibrous composite (TFNC) membrane. After fine-tuning of membrane preparation process, the resultant membrane exhibited competitive PV performance for dehydration of 90 wt% isopropanol solution. The unique three-tier composite membrane suggested an effective and facile approach to design novel membrane structure for further improvement of PV performance.

**KEYNOTE SPEECH**

# Carbon nanotube-related thermoelectric composites and flexible devices

Guangming Chen
Shenzhen University

In recent years, organic/inorganic thermoelectric (TE) composites and the corresponding flexible devices have witnessed significant achievements and become a hotspot in the fields of energy and materials. Here, we introduce some recent research results of the carbon nanotube (CNT)-related TE composites and their flexible devices in our group.

1) By "template-directed in situ polymerization" method, we prepared poly(3, 4-ethylenedioxythiophene)/multi-walled CNT (PEDOT/MWCNT) and polypyrrole (PPy)/MWCNT composites. In the composites, a simple cable-like morphology is clearly observed.

2) The TE performances are greatly improved and can be conveniently adjusted via nanostructure fabrication of polymer (PEDOT or PPy) nanowire/single-walled CNT (SWCNT) composites. Three typical TE composites with 3D interconnected networks, layered structure, and unusual lychee-like morphology have been achieved.

3) Recently, a 3-phase electropolymerization method has been employed in the fabrication of PEDOT/SWCNT TE composites, which display high electrical conductivity and enhanced power factor.

4) Non-conducting polymers (poly-Schiff-bases) have been employed in fabrication of PEDOT TE composites. In addition, ternary TE composites of PEDOT/graphene/SWCNT, have been studied as well.

5) A series of n-type and p-type TE composites of SWNCT with small organic molecules have been successfully prepared. NDINE or PDINE, DETA reduction followed by $CaH_2$ treatment, and alkylammonium cationic surfactants have been employed in the fabrication of n-type SWCNT-based TE composites. Additionally, a p-type TE composite of SWCNT using narrow-bandgap small organic molecule (TPNO) coating has been obtained.

6) Aerogels of graphene/SWCNT and hydrogels of PMMA/SWCNT composites have been achieved, which reveal attracting TE performances.

7) Flexible films and devices of SWCNT-based TE composites have been fabricated. The output properties follow the order of serial > folding > stacking styles for the devices using flexible films. Moreover, we put forward a method to evaluate flexible device assembly modes: , where FDP, Pmax, m, $\Delta T$ and N stand for flexible device performance, maximum output power, product mass, temperature gradient and pair number of p-n couples, respectively.

Acknowledgement:
The financial support by National Natural Science Foundation of China (No. 51573190) is acknowledged.

INVITED LECTURE

# A facile strategy for fabricating the hierarchical nanocomposites of 1D V2O5 nanowire on 3D N-doped carbon materials with synergistic effect for supercapacitor

Wei Sun[1, 2], Guangming Wu[2], Zhengwei You[1]
1.Donghua University
2.Tongji University

As one of the most promising candidates for supercapcitor electrodes, vanadium pentoxide (V2O5) usually suffers from the poor electronic conductivity and few accessible active sites. The key to solve these problems is to improve the conductivity of V2O5 by integrating conductive carbon materials and constructing rational architectures. Here we introduce a facile strategy to construct hierarchical nanocomposites by combing 1D V2O5 nanostructures with 3D N-doped carbon porous networks. The V2O5 nanowires are aligned vertically on porous conductive carbon backbone or uniformly coated on surface of carbon nanofibers to form 3D porous networks with core/shell nanostructures. We find that the introduction of N atoms into carbon lattice plays an import role in the formation of veritical VOx nanowires arrays or core/shell structures on carbon materials, which is attributed to different VOx nucleation processes. First-principles simulations and the relative experiments are managed to demonstrate the effect of incorporating N atoms into conductive carbon backbone on the formation of different V2O5 nanostructures on it. Such a design offers distinct advantages for V2O5-based materials for supercapacitors. As a consequence, it delivers an ultrahigh specific capacitance of 710 F g-1 (at current density of 0.5 A g-1) and exhibits outstanding rate performance and good cycling behavior (after 20000 cycles 95 % retention of specific capacitance). This study will open a way to guide the fabrication of nanocomposites by combining different dimensional nanomaterials.

ORAL PRESENTATION

# A scalable, flexible and ultralight hybrid supercapacitor system design and implementation using renewable materials

Bhupender Pal[1], Shengyuan Yang[1, 2], Mashitah Mohd. Yusoff[1], Rajan Jose[1]
1.Universiti Malaysia Pahang
2.State Key Laboratory for Modification of Chemical Fibers and Polymer Materials， International Joint Laboratory for Advanced Fiber and Low-dimension Materials， College of Materials Science and Engineering， Donghua University， Shanghai 201620， P. R. China

The affordable and abundant demand for high performance and flexible energy-storage devices have been promoted by the recent advancement in wearable and portable electronics.1, 2 In this research, we have developed scalable, cost-effective, eco-friendly and easily-operated method to fabricate cellulose/reduced graphene oxide based hybrid nanofibers for supercapacitor electrodes by electrospinning technique. These binder-free and free-standing electrodes can serve as current collector and working electrode due to the good electrical conductivity, thus simplifying the assembly of device and reducing the cost of supercapacitors. The cellulose has been synthesized from natural waste/bio based materials and also have used as a separator and substrate for active electrode materials (cathode and anode). The same cellulose has been further used for the preparation of porous carbon and graphene by carbonization and wet chemical method technique. The cellulose as a current collector can effectively absorb electrolyte and act as electrolyte reservoirs to facilitate ion transport. As-synthesized materials have been characterized by various techniques such as FT-IR, XPS, XRD, FE-SEM etc. for their identification. After the synthesis of all materials, three electrode configuration systems have been measured for the optimization of electrochemical properties, then two electrode systems (prototype device) have been fabricate by using cellulose/reduced graphene oxide hybrid paper electrode. The electrochemical properties and electrochemical stability of the prototype device have been tested by cyclic voltammetry, Galvanostatic charge/discharge and electrochemical impedance spectroscopy.

References

1. C. Zhong, Y. Deng, W. Hu, J. Qiao, L. Zhang and J. Zhang, Chem. Soc. Rev., 2015, 44, 7484–7539.
2. P. G. Bruce, S. A. Freunberger, L. J. Hardwick and J. M. Tarascon, Nat. Mater., 2012, 11, 172.

POSTER PRESENTATION

# Synthesis of novel imine-linked metal porphyrin conjugated microporous polymer with enhanced visible-light photocatalytic activity

Xu Cui, Yanhui Li, Qian Duan
Changchun University of Science and Technology

The use of visible light to drive degradation organic dyes has attracted enormous attention, due to the straightforward, mild and environmentally friendship process as well as a sustainable way for efficient water purification and remediation technologies. In this work, a metal porphyrin-based conjugated microporous polymers NiTAPP-CMP was easily prepared with the 5, 10, 15, 20-tetrakis(para-aminophenyl) porphyrin Nickel (II) (NiTAPP) and 2, 6-dimethoxynaphthalene- 1, 5-dicarbaldehyde (DHNDA) via aldehyde-amine condensation reaction under the solvothermal condition and used for the photodegradation of organic pollutants. As a result, the as-obtained NiTAPP-CMP exhibit excellent photocatalytic efficiency, recyclability, and stability. These results indicate that the NiTAPP-CMP is potentially applicable in the environmental purification of organic pollutants in the industrial wastewater.

POSTER PRESENTATION

# Highly efficient visible-light-driven Nitrogen-Doped Porous Carbon-ZnO nanocomposite photocatalysts

Xiao Chen, Jian Cui, Xuran Xu , Jieshu Qian, Dongping Sun
Nanjing University of Science and Technology

Reliable production of photocatalyst with tunable structure and property is of great importance for the elimination of organic pollutants in water. However, the directed band gap limites the light absorption to sunlight, therefore restricting the photocatalytic efficiency of ZnO-based photocatalysts. Herein, an efficient photoactive material, nitrogen-doped porous carbon-ZnO (N-ZnO/CBC) was prepared by direct carbonization of zeolitic imidazolate framework (ZIF)-8 using network structural bacterial cellulose(BC) as substrate in a nitrogen atmosphere. Results N-ZnO/CBC650 showed the best photocatalytic activity on tetracycline hydrochloride and dichlorophenol degradation, due to the carbon/nitrogen dopants and oxygen vacancies. The mechanism was illustrated and showed that hydroxyl radicals and photo-excited electrons contributed to the dye degradation.With more virtues of excellent adsorption performance, reusability and stability, our sample has shown its great potential for the removal of organic molecules in real polluted water.

**POSTER PRESENTATION**

# Two-layer fibre mats fabricated from electrospun fibres covered on rotating spun fibres for high-efficiency air filtration

Yongzhen Li1, Puwang Li1, Mingzhe Lv1, Ziming Yang1, Guiang Liu3, Lingxue Kong1
1.Chinese Academy of Tropical Agriculture Sciences
2.Deakin University
3.Lingnan Normal University

Particulate matter (PM) pollution has cuased healthy issue and raised much techinical and political concern recent years in developing China. The existing air filters meet contradictory issue in combining removal efficiency and air permeability, which refers to tiny structure and strong mechanical properties of filtering materials. Air filtering materials play two main roles: separating aerosol particles from air flow and removing bacterial. Air pollution was caused by toxic dust particles and bacteria floating in the air. Micro-/nano fibres with large surface areas are potentially used for air filtering in expectation of high filtering efficiency. Fine fibres made from rotating spinning of PVA, PAN and PS were investigated for air filtering in terms of fibres' physical properties and morphologies. Rotating spinning showed competitive power in production yield. Porous fibre mats could filter particles which were larger than the pores in fibre mats. However, lower filtering efficiency of PM2.5(<85%) was observed in rotating spun fibres compared with electrospun fibres (>95%). To enhance filtering efficency , electrospun fibres were collected on rotating fibres for fabrication of two-layer filtering material. The combination of nano and microstructure fibres would get fibre web with higher mechanical properties similar to microfibers and filtering efficiency as electrospun nanofibers web (more than 95% for PM 2.5). In conclusion, the two-layer fibre combination of electrospun nanofibers and rotating spun microfibers could get strong filtering materials with high efficiency for air filtering.

POSTER PRESENTATION

# Temperature and $CO_2$ dual-responsive pickering emulsions using Jeffamine M2005-modified cellulose nanocrystals

Gaihuan Ren1, Jianxin He1, Yuman Zhou1, Xiaoyan Sun1, Dejun Sun2
1.Zhongyuan University of Technology
2.Shandong University

Cellulose nanocrystals (CNCs) with good biodegradability are promising biomaterial for responsive Pickering emulsifiers. However, the high hydrophilicity prevents the emulsification ability. Some existing studies utilize complicated covalent synthesis to enhance the emulsification ability of CNCs. Our study focuses on CNCs efficient hydrophobization via simple and controllable electrostatic interaction by using thermosensitive Jeffamine M2005. The obtained partially hydrophobic CNCs-M2005 possessed thermo and CO2 dual-responsive properties. At 20 °C, partially hydrophobic CNCs-M2005 nanoparticles were adsorbed on the oil-water interface and stabilize the O/W Pickering emulsion. After heating to 60 °C for 20 min, the emulsion became oil-water separated. The temperature-induced demulsification was resulted from the dehydration of PEO and PPO caused aggregation of CNCs-M2005, which has been proved by DLS and TEM experiments. Besides, stable emulsion became oil-water separated after bubbling CO2, which was attributed to the dissociation of partially hydrophobic CNCs-M2005 into CNCs-COOH and M2005+ in acidic environment. The thermo and CO2 dual-responsive bio-safe Pickering emulsion paves the way for design intelligent food, cosmetics, and drug delivery systems.

POSTER PRESENTATION

# PVDF/$Fe_3O_4$ composite nanofiber membrane with magnetic effect and coupled electret effect and its application in air filtration

Fan Liu, Mengying Li, Xiang Li, Weili Shao, Jianxin He
Textile college of Zhongyuan University of Technology

As air pollution has become a major environmental concern, it has become urgent to develop anti-haze filtration material to protect the public effectively. In this study, a PVDF/Fe3O4 electret nanofiber membrane filter material was prepared using electrospinning technology. The average diameter of the fiber was approximately 243 nm. Fe3O4 was prepared by a coprecipitation method. When the temperature of the water bath was 60 °C, the particle size of the nanoparticles reached approximately 15 nm, and the distribution was uniform. The mechanical properties of the PVDF/Fe3O4 composite nanofiber membrane reached 6.32 MPa. The discharge peak of PVDF/Fe3O4 appeared at 122 °C while maintaining a high surface potential of 2.08 kV after seven days. PVDF/Fe3O4 can maintain an excellent filtration efficiency of 99.95% and a low-resistance pressure drop of 58.5 Pa at a density of 2.06 g/m2 and airflow of 32 L/min. Therefore, PVDF/Fe3O4 composite nanofiber membranes have potential applications in the field of air filtration.

**POSTER PRESENTATION**

# Cone-like Titanate immobilized on PAN nanofibers: hierarchical architecture for effective photocatalytic property

Mingyue Zhu, Xiaofei Guo, Si Cheng, Lixing Dai, Chuanxiang Qin
Soochow University

Hybrid polyacrylonitrile nanofibers (PAN NF) is widely employed in environmental remediation. A new composite photocatalyst, based on flexible functional polyacrylonitrile nanofibers (f-PAN NF), was developed through depositing composite coatings of α-$TiO_2$ and cone-like titanate ($H_2Ti_5O_{11}\cdot 3H_2O$) successively. There were much hydroxyl groups on surface of α-$TiO_2$, so it attached to f-PAN NF tightly by possible chemical bond. α-$TiO_2$ worked as seed and accelerated the nucleation of titanate in the following process. During the deposition process of titanate, smaller-sized nanoparticles which gradually appeared in precursor solution dissolved and re-deposited on larger-sized nanoparticles due to their higher surface energy.[1] Besides, cyanuric acid as morphology controller contributed to the cone-like structure of titanate. Based on "Ostwald Ripening" and the assist of cyanuric acid, cone-like titanate could deposite on α-$TiO_2$@f-PAN NF uniformly and tightly at 35 °C. Due to the relative low reaction temperature, abundant surface hydroxyl groups of titanate were retained.[2] These hydroxyl groups could act as capture centers for photo generated electrons and absorb more pollution to promote photocatalytic activity.[3] Owing to these active groups and uniform distribution of titanate, the photocatalytic performance of hybrid f-PAN NF was remarkable under LED light irradiation. In addition, the composite photocatalyst exhibited stable reusability and recollection because of the special design for attachment between nanofibers and coatings.

References:

1. H.G. Yang, H.C. Zeng, The Journal of Physical Chemistry B, 2004, 108, 3492-3495.
2. Y. Xu, W. Wen, J.M. Wu, Journal of Hazardous Materials, 2018, 34, 285-297.
3. S.H. Szczepankiewicz, A.J. Colussi, The Journal of Physical Chemistry B, 2000, 104, 9842-9850.

Acknowledgements
This report is supported by the Natural Science Foundation of Jiangsu Province (Grant No. BK20161214).

POSTER PRESENTATION

# Degradation performance experiment of paper and straw-based seedling pot

Xizhi Jiang, Lei Xu, Nina Yan, Rong Zhang, Min Feng
Jiangsu Academy of Agricultural Sciences

Most of the articles about the degradation performance of biodegradable nursery pots were focused on degradation morphology, strength of pot body, changes of soil, effect of raising seedings and so on. However, it rarely study on degradation rate, molecular structure of degradation process and changes of crystallinity. In order to study the degradation performance of biodegradable nursery pots, three types containing paper pot, undecomposed straw pot and decomposed straw pot were used to carry out the experiment in this paper. The degradation rate, molecular structure and crystallinity were evaluated and analyzed through the experiments of degradation test in soil, Fourier infrared scanning and x-ray diffraction. It is necessary to have a further research on degradation performance of biodegradable nursery containers and mechanism of degradation. The results showed that the degradation rate of paper pot with the best degradation property was growing at a rate of about 2 times. The maximum degradation rate was up to 85%. The undecomposed straw pot has always maintained a stable and increasing degradation rate. The undecomposed straw pot with degradation rate of 78% in the eighth week was higher than that of decomposed straw pot. The changes of mass loss and degradation rate were likely to be affect by a variety of microbial community structure. On the one hand, due to partial degradation of cellulose and semicellulose, the remaining part kept stable performance. On the other hand, the changes of microbial structure and activity affected the degradation performance. By comparison, FTIR absorption strength of the biodegradable nursery pots was decreased in different degree due to the changes of molecular structure during biodegradation. The absorption strength of paper pot decreased in the location of 3340 $cm^{-1}$ and 2920 $cm^{-1}$ indicated that the degradation of cellulose and hemicellulose. In addition, the relative peak of 1640 $cm^{-1}$ and 1460 $cm^{-1}$ decreased with the further degradation, showing that lignin was degraded in the biological reaction. The peak strength of CH stretch band was enhanced obviously at 892 $cm^{-1}$. It showed that the substitution reaction of aromatic ring was an important part of the degradation process of lignin. The main group structure of undecomposed straw pot and decomposed straw pot were basically the same. The absorption peaks strength decreased at 3340 $cm^{-1}$ and 1010 $cm^{-1}$ might be due to the oxidative decomposition of benzene methanol unit. The peak strength of CH stretch band decreased at 2920 $cm^{-1}$ probably because of the decomposition of carbohydrates and the fracture of partial fatty chain. The degradation of lignin resulted in the decrease of peak strength of 1650 $cm^{-1}$. The x-ray diffraction peak intensity of paper pot with the best degradation property decreased obviously after 8 weeks test. The diffraction peak disappeared at $2\theta=15.6°$. At the same time, The intensity of diffraction peak decreased at $2\theta=22.5°$. In addition, it appeared new diffraction peaks at $2\theta=27°$ of undecomposed straw pot and decomposed straw pot. This might be due to the changes of crystalline which caused by the degradation of straw. The biodegradable nursery containers with excellent degradation property could improve the problem of environmental pollution effectively.

POSTER PRESENTATION

# Facile preparation of PAM/CS/$Fe_3O_4$ composite hydrogels for effective removal of methylene blue from aqueous solution

Cheng Zhang, Xiaoqian Wen, Yinqiu Wu, Zheng Cao, Junfeng Cheng, Chunlin Liu
Changzhou University

In this study, Fe3O4 magnetic nanoparticles were synthesized in situ in the polyacrylamide/chitosan (PAAm/CS) hydrogel networks, and the PAAm/CS/Fe3O4 magnetic hydrogel was prepared and characterized by Scanning electron microscopy, X-ray diffraction, and Raman spectroscopy. The results confirmed that the three-dimensional network structure of the hydrogel was loaded with Fe3O4 nanoparticles. The adsorption property of PAAm/CS and PAAm/CS/Fe3O4 hydrogels for methylene blue (MB) in aqueous solution was studied by UV-Vis spectrometer. The results indicated that compared to the PAAm/CS hydrogel, the PAAm/CS/Fe3O4 hydrogels can adsorb MB with a higher adsorption capacity of approximately 1603 mg/g, and the kinetics and isotherm models of the adsorption process could be better described by the pseudo-first order model and Langmuir isotherm model, respectively. The PAAm/CS/Fe3O4 hydrogel showed good adsorption performance for MB and great potential in removing water contaminants.

POSTER PRESENTATION

# Modified HEC/PAA complex membranes for water-in-oil emulsion separation

Liping Zhu, Dafaalla Babiker, Shuguang Yang
Donghua University

Oil-water separation has recently become a global challenge due to the increased production of industrial oily wastewater and frequent oil spill accidents during oil production and marine transportation. The growing environmental concerns and economic demands drive the needs to develop effective solutions for oil-water separation, especially for the oil/water emulsions. Membrane separation technology has gradually become an important and promising method in oil water separation field due to its high efficiency, low energy consumption and green characteristics. In this work, a highly efficient membrane with low-cost is prepared with green materials via a simple pathway for separation of oil from water-in-oil emulsion. Hydroxyethyl cellulose (HEC) and poly(acrylic acid) (PAA) are selected to form a complex membrane on the surface of poly(ethylene terephthalate) (PET) nonwoven fabric via a layer-by-layer (LbL) assembled method. The HEC/PAA membrane is then modified through a thermal or chemical induced crosslinking treatment to obtain a hydrophobic surface. The structure and properties of HEC/PAA coated PET membrane before and after the crosslinking treatment are investigated and compared through FE-SEM, FT-IR, XPS and contact angle measurement. The modified membranes are then applied as oil filters to treat water-in-oil emulsions with different concentrations. Both of the modified membranes show excellent separation efficiencies with a more than 99.4% rejection for all tested water-in-oil emulsions.

**POSTER PRESENTATION**

# Enhancing pervaporation dehydration performance by manipulating the interfacial interactions via introducing polyethyleneimine modified graphene oxide intermediate layer

Peiyun Li, Xuefen Wang
Donghua University

A novel three-tier composite membrane based on highly porous nanofibrous substrate was demonstrated for efficient isopropanol dehydration by pervaporation. Polyethyleneimine (PEI) modified graphene oxide (GO) sheets were vacuum-assistant assembled onto porous electrospun polyacrylonitrile (PAN) nanofibrous substrate to achieve a smooth, hydrophilic and compact PEI-GO intermediate layer. The introduction of PEI chains endowed GO interlayer with sufficient interaction for bonding adjacent GO nanosheets to enhance stability in water/isopropanol mixture and also with the ascended interlamellar space to improve the water-sorption ability due to the abundant active amino groups. Benefiting from PEI-GO layer, a defect-free sodium alginate (SA) skin layer could be facilely manufactured with elaborately controlled thickness as thin as possible in order to reduce mass transfer resistant and enhance permeability maximally. And the interlayer contributed to enhance interfacial adhesion to promote the structure integrity of three-tier thin-film nanofibrous composite (TFNC) membrane. After fine-tuning of membrane preparation process, the resultant membrane exhibited competitive PV performance for dehydration of 90 wt% isopropanol solution. The unique three-tier composite membrane suggested an effective and facile approach to design novel membrane structure for further improvement of PV performance.

POSTER PRESENTATION

# Wearable wool knitwear with shape memory effect

Mohammad Irfan Iqbal1, Jinlian Hu1
1.Institute of Textiles and Clothing， The Hong Kong Polytechnic University
2.The Hong Kong

Investigation into wool knitwear along with shape memory ability has great potential in order to rediscover the wool as a smart apparel. The aim of this study is to explore the shape memory ability of 100% wool-based knitwear stimulated with water. This study presents the findings of shape memory ability of wool in the form of fibers, yarns and fabrics using optical camera, light microscope and image analysis software. The water stimulated fabric has shown 20% more area change compared to dry sample. Moreover, the effect of fabric structure on shape memory performance of wool knitwear has been analyzed and discussed. It has been confirmed that regardless of structure the wool knitwear has shape memory property. These findings provide fresh insight into developing sustainable smart textiles using natural fibers like wool. This material can give the similar response upon contact with body sweat and the humid environment.

POSTER PRESENTATION

# $Fe_3O_4$@CNF prepared from cellulose derivative and its application in Li-ion battery

Yumei Gong1, Weihao Han1, Yunfei Chang1, Qinghua Zhang2
1.School of Textile and Material Engineering, Dalian Polytechnic University
2.The State Key Laboratory for Modification of Chemical Fibers and Polymer Materials, College of Material Science and Engineering, Donghua University

Rechargeable Li-ion battery (LIB) has been considered as one of the most promising energy storage systems due to its long cycle life, high energy density, high rate capability and environment-friendliness. Fe3O4@CNF anode material for LIB was synthesized by using deacetylated cellulose as the carbon nanofiber (CNF) source and Fe(acac)3 as the Fe3O4 source through electrospinning method in this report. The CNF buffer the volume change of Fe3O4 during electrochemical cycling and improve the electrical conductivity. Meanwhile, introduction Fe3O4 to CNF provided a larger specific surface area and more mesopores which promoted electrolyte penetration and lithium ions diffusion. As anode materials for LIB, Fe3O4@CNF exhibits more stable cycling performance and high rate performance than FexOy nanofibers without a carbon matrix (FexOyNF). Fe3O4@CNF electrode demonstrates high reversible capacities of 773.6 and 596.5 mA h·g-1 after 300 cycles at a large current densities of 1 and 2 A·g-1 and the capacity retentions measured from the second cycle were 95.0% and 109.5%, respectively.

POSTER PRESENTATION

# Morphology dependent capacitive performance for nanostructured polyaniline fabricated through ice–ice interfacial polymerization in fully aqueous solution

Lifeng Yang, Zongyi Qin, Yueying Shen, Shuo Hu
Donghua University

In this work, we propose a simple and environmentally friendly method to fabricate polyaniline, which called ice–ice interfacial polymerization. At first, aniline and ammonium persulphate (APS) were dissolved in 1 M HCl aqueous solution, respectively. And then, just as the name implied, aniline was firstly frozen at -18℃. After 12 hours, APS was added to the breaker where aniline had become ice. Finally, the whole system was cooled at -18℃ for 12 hours. The influences of the interface area on the morphologies and micro-structure were investigated by SEM, FT-IR and XRD. Furthermore, the electrochemical properties were evaluated by electrochemical workstation in 1 M $H_2SO_4$ solution under a three-electrode system. Polyaniline with coral-like structure were both obtained under various conditions. But polyaniline prepared in a beaker (250 ml, ~7.6 cm in diameter) exhibited the largest specific capacitance of 432.2 F/g. In the case of the same volume, a larger interface area means a lower height, which will speed up the reaction. In summary, this work provides a new way to fabricate polyaniline for supercapacitor applications.

POSTER PRESENTATION

# 无融缩抗氧化聚苯硫醚砜纤维的制备及其性能研究

王明稳，李振环，苏坤梅
天津工业大学

在不同温度下以双氧水为氧化剂、冰乙酸为催化剂对聚苯硫醚（PPS）纤维进行氧化改性处理。通过改变反应温度和反应时间等制备工艺，探究聚苯硫醚砜（简称 PPSO，该产品中含有-S-、-SO-和-$SO_2$-）纤维的形貌及其性能。利用热重分析法（TG）、差示扫描量热法（DSC）、XPS、SEM、力学性能测试等对改性纤维结构和性能进行了表征。结果表明：当双氧水、蒸馏水、冰乙酸质量比为 50：25:25，处理时间为 5 个小时时，能够得到具有高强度、耐高温、抗氧化的 PPSO 纤维；氧化改性过程中，伴随着 S 原子的流失和聚芳烃的生成，PPSO 纤维中氧含量大幅增加，证明 PPS 纤维被成功氧化改性；PPSO 纤维力学强度仅次于 PPS 纤维并在 300℃处理 2 个小时后纤维力学强度保持率高达 80%，这说明 PPSO 纤维具有高强度和耐高温等特性；此外，改性后的 PPSO 纤维经硝酸浸泡处理后强度保持率显著大于 PPS 纤维，说明改性后的 PPSO 纤维抗氧化能力明显提高。

POSTER PRESENTATION

# 液面成膜法制备嵌段共聚物超薄复合分离膜

严旋娜[1,2]，汪勇[2]
1.Jiangsu Academy of Agricultural Sciences
2.Nanjing Tech University

在面向水处理的超滤膜中，嵌段共聚物分离膜因为具有孔径分布窄、孔隙率高、物化性质可调的多孔结构而被认为是最有前景的分离材料之一。然而，如何有效制备超薄嵌段共聚物复合膜依然是一大挑战。我们通过设计采用液面成膜的方法制备了基于嵌段共聚物选择性溶胀的复合超滤膜。将不与水互溶的聚苯乙烯-*block*-聚 2-乙烯基吡啶溶液滴加于水面，水的高表面张力将使溶液在其表面均匀铺展。之后随着溶液的挥发，在水面会逐渐形成一层嵌段共聚物薄层。将其与大孔聚偏氟乙烯（PVDF）基膜复合并经过热处理及选择性溶胀后，得到可用于分离的、分离层厚度仅为~240nm 的嵌段共聚物复合膜。大孔 PVDF 基膜的引入保证了复合膜的机械强度。而超薄的嵌段共聚物膜层使溶胀后复合膜的渗透性能显著优于商业化超滤膜。经测试，在 55℃溶胀 5 h 时，复合膜的纯水渗透性可达 955.7 L $m^{-2}$ $h^{-1}$ $bar^{-1}$，而此时对 7nm 牛血清蛋白的截留率依然高达 77.5%。复合膜的渗透分离性能也可以简单地通过溶胀时间的改变进行调控。该方法在降低制膜难度的同时，也为大面积超薄膜层的制备提供了可能。

**INVITED LECTURE**

# "Mist polymerization" method for fabricating superhydrophobic cotton fabrics

Xiangdong Liu
School of Material Science and Engineering, Zhejiang Sci-Tech University

Recently, our group developed a surface modification technology using "mist polymerization" to fabricate functional cotton fabrics. The "mist polymerization" method have both the advantages of liquid-phase and gas-phase reactions: less damages on the structure of the matrix material, very fine and controllable layer, wide applicable monomer range, and small monomer dosage. Here, I summarize recent works on the application of the "mist polymerization" method to fabricate superhydrophobic fabrics.

INVITED LECTURE

# Energy harvesting and storage devices for E-textiles and stretchable electronics

Xiong Pu
1. Beijing Institute of Nanoenergy and Nanosystems Chinese Academy of Sciences
2. University of Chinese Academy of Sciences, School of Nano Science and Technology

Rapid advancements in wearable smart textiles and stretchable electronics impose the challenge on corresponding power devices that they should possess comparable flexibility, stretchability and functionalities. In this talk, we briefly summarize our recent progresses on wearable energy storage and harvesting devices, especially energies devices aiming to power smart textiles and artificial electronic skins. Both yarn or textile-based electrochemical energy storage devices and textile-based triboelectric nanogenerators (TENGs) have been developed; meantime, self-charging systems have been achieved by integrating them so that energies harvested from daily human motions can be stored simultaneously. It is even more challenging to provide power sources for electronic skins or soft electronics/robotics. We report a soft skin-like triboelectric nanogenerator that enables both biomechanical energy harvesting and tactile sensing by hybridizing elastomer and ionic hydrogel as the electrification layer and electrode, respectively. Ultra-high stretchability and transparency are achieved simultaneously for an energy-harvesting device. Lastly, we report our updated progresses on and self-healable soft TENGs by utilizing self-healable elastomers. Bulked thin layers of conductive polymers are achieved by transfer the film onto pre-strained self-healable elastomers, which later ensure the realization of self-healing, stretchability, and transparency of the TENG simultaneously. Our work provides new opportunities for soft power sources and potential applications in soft/wearable electronics.

**INVITED LECTURE**

# Ultralarge mode area chalcogenide photonic crystal fiber for 100-W high-power mid-infrared applications

Xian Feng, He Ren, Sisheng Qi, Yongsheng Hu, Feng Han, Jindan Shi, Zhiyong Yang
Jiangsu Normal University

We report our recent progress on the development of mid-infrared chalcogenide glass photonic crystal fiber (PCF). Ge-As-S-Se chalcogenide glass system is chosen for constructing the all-solid PCF structure. Broadband single-mode chalcogenide glass photonic crystal fiber, i.e., the 'endless' single-mode photonic crystal fiber, with ultralarge large mode area (ULMA) > 5000 μm2 has been designed and fabricated. Single-mode operation has been experimentally demonstrated in the mid-infrared region. Broadband mid-infrared supercontinuum has been generated in the chalcogenide ULMA PCF, pumped by an ultrafast femtosecond longwave laser. The emerging chalcogenide ULMA PCF technology opens the way towards mid-infrared applications requiring 100W-level high power.

INVITED LECTURE

# Advanced functional semiconductor fibers

Lei Wei
Nanyang Technological University

We present the recent progress on laser-induced directional crystallization to enable the precise control on crystal structures and device density of in-fiber semiconductor devices. These fibers are particularly suitable for wearable electronics for full-body sensing.

**INVITED LECTURE**

# Application of polymer fibers for supported catalysis in feasible practice

Xian-Lei Shi
Henan Polytechnic University

The earlier a novel catalyst could be certified with high-efficiency in the feasible practice, the more possibilities and ways to put it into real industrial productions. In recent studies of our group, the synthetic fibers such as polyacrylonitrile (PAN), polypropylene fiber (PP), polybenzimidazoles (PBI) and polyphenylene sulfide (PPS), polyether ether ketone (PEEK) etc. were served as the supports to prepare fiber-supported base or acid catalysts, fiber-supported PTC, fiber-supported ionic liquids and fiber-supported metal active sites, which were set in the continuous-flow processing or spinning basket reactor for heterogeneous catalysis and verified in a series of reactions ranging from organic synthesis to biomass and CO2 conversions, [1-10] and were deemed to be one of the most promising catalyst forms from the reaction engineering point of view. Moreover, this newly developed fiber-based catalytic method provides a new strategy to design and develop greener, safer and more economical processes in the chemical industry.

References

[1] X.-L. Shi, B. Sun, Y. Chen, et al., Green. Chem, 2019, 21, 3573-3582.
[2] X.-L. Shi, B. Sun, Y. Chen, et al., J. Catal. 2019, 372, 321-329.
[3] X.-L. Shi, Q. Hu, Y. Chen, et al., J. Ind. Eng. Chem, 2019, 69, 387-396.
[4] X.-L. Shi, Y. Chen, P. Duan, et al., ACS Sustainable Chem. Eng. 2018, 6, 7119-7127.
[5] X.-L. Shi, Y. Chen, Q. Hu, et al., J. Ind. Eng. Chem. 2018, 60, 333-340.
[6] X.-L. Shi, Y. Chen, Q. Hu, et al., Ind. Eng. Chem. Res. 2018, 57, 7450-7457.
[7] X.-L. Shi, Y. Chen, Q. Hu, et al., J. Ind. Eng. Chem. 2017, 53, 134-142.
[8] X.-L. Shi, Q. Hu, F. Wang, et al., J. Catal. 2016, 337, 233-239.
[9] X.-L. Shi, X. Xing, H. Lin, et al., Adv. Synth. Catal. 2014, 356, 2349-2354.
[10] X.-L. Shi, M. Zhang, Y. Li, et al., Green Chem. 2013, 15, 3438-3445.

INVITED LECTURE

# Polymer complexation for adaptive fibers

Shuguang Yang*, Dezhong Liu, Wentao Huang, Zexin Liu

State Key Laboratory for Modification of Chemical Fibers and Polymer Materials, Center for Advanced Low-dimension Materials, College of Materials Science and Engineering, Donghua University

Polymer complex is a polymer association of different polymers which have relatively strong intermolecular interactions or stereo match effect between polymer chains. Polymer complexes have been utilized in various fields, such as food, cosmetic, and pharmacy and biomedical materials. Fiber is one of the most important material forms, and shaping polymer complex into fiber will further extend its applications. We have developed different methods utilizing the polymer complexation to produce fibers. Owing to supramolecular interactions among the polymers complex, the fiber exhibited various adaptive behavior, such as mechanical, humidity, and electrical responsive. In this lecture, we will introduce our research on how use polymer complexation to prepare fibers and investigate their adaptive behaviors

Reference

1. J. Li, Z. Wang, L. Wen, J. Nie, S. Yang, J. Xu, and S. Z. D. Cheng, ACS Macro Lett. 2016, 5, 814–818.
2. J. Nie, Z. Wang, J. Li, Y. Gong, J. Sun, S. Yang, Chinese Journal of Polymer Science, 2017, 35, 1001–1008.
3. D. Liu, J. Li, W. Huang, S. Yang, Acta Polymerica Sinica 2018, 445-455.
4. J. Li, W. Huang, D. Liu, J. Sun, L. Zhu, Y. Wang, S. Yang, Materials Today Communications 2019, 19, 425-432.
5. W. Huang, D. Liu, J. Li, L. Zhu, S. Yang, Science China Technological Sciences 2019, 62, 931-944.
6. J. Li, J. Sun,·D. Wu, W. Huang, M. Zhu, E. Reichmanis, S. Yang, Advanced Fiber Materials 2019, 1, 71–81.

**INVITED LECTURE**

# A compact tactile sensor based on optical micro/nanofiber for tissue stiffness detection

Yao Tang, Jing Pan, Zhang Zhang, Yue Xu, Lei Zhang, Liming Tong
College of optical science and engineering，Zhejiang University

With the advance of robotic surgery, doctors could perform many types of complex procedures with more precision, flexibility and control than is possible with conventional techniques. However, the mechanical properties of the target organ, which are critical determinants of function in oncogenic transformation and hepatic fibrosis, are not sufficient to the physician because the doctor can not touch the organ directly through the tiny incisions of the minimally invasive surgery. Although a few commercial surgical robots have mounted tactile sensors, it remains a great challenge for the surgical robots to sense the mechanical properties of a specific tissue with high sensitivity and high resolution. In this work, we demonstrate a compact tactile sensor based on optical micro/nanofiber (MNF) for tissue stiffness detection. A biconical tapered fiber was fabricated by stretching a standard optical fiber while heating it with a hydrogen/oxygen flame. The as-fabricated biconical tapered fiber was attached onto a silica capillary with the taper waist (i.e., MNF) located at one end of the silica capillary. The MNF and silica capillary were then capsulated by polydimethylsiloxane, forming a compact and robust probe with an outer diameter of ~1.5 mm, which can be inserted into the operating channel of an endoscope (usually 2-4 mm in diameter). When the probe touches different materials with the same pressure, the flexible probe undergoes a deformation which is a function of the stiffness of the materials. By measuring the transmission of the MNF, highly sensitive tissue stiffness detection is realized.

INVITED LECTURE

# Smart patterned polymer surface with dynamic wrinkles

Xuesong Jiang
Shanghai Jiaotong University

The responsive micro/nanoscale patterns on the surface whose morphology can be tuned dynamically by environmental stimulus can possibly enable the on-demand control of the surface properties, and provide an important alternative to realize the smart surface. Recently, we developed a facile and effective strategy for the fabrication of a reversible pattern with a morphology that can be dynamically erased and tuned in-situ by light, pH and temperature through introducing the dynamic chemistry into wrinkling systems. Taking dynamic photo-dimerization of anthracene as example, the key point for this strategy is that the modulus of the top layer can be tuned by the reversible cross-linking via photo-dimerization. The reversible nature of the photo-dimerization of anthracene enables the dynamic change of the pattern morphology from the smooth state to the wrinkle pattern, allowing for accurate control of the adhesion, wettability and optical properties of the resulting surface. This complex tunable wrinkle patterns provides an alternative platform to develep smart optical and photonic devices.

**INVITED LECTURE**

# Soft electronic devices using flexible, stretchable and mendable polymeric materials

Wei Lin Leong
Nanyang Technological University

Organic electrochemical transistors (OECTs) are highly attractive for applications ranging from circuit elements, neuromorphic devices to transducers for biological sensing and the archetypal channel material is poly(3, 4-ethylenedioxythiophene):poly(styrenesulfonate), PEDOT:PSS. The operation of OECTs involves the doping and de-doping of a conjugated polymer due to ion intercalation under the application of a gate voltage. However, the challenge is the trade-off in morphology for mixed conduction since good electronic charge transport requires a high degree of ordering among PEDOT chains, while efficient ion uptake and volumetric doping necessitates open and loose packing of the polymer chains. Here we demonstrate ionic liquid doped PEDOT:PSS that overcomes this limitation. Ionic liquid doped OECTs show high transconductance, fast transient response and high device stability over 3600 switching cycles. The OECTs are further capable of having good ion-sensitivity. We also look at strategies to achieve polymer films which are robust towards physical deformation and able to self-heal without compromising electronic properties as well as deposition routes for making devices on unconventional curved substrates. Further, we demonstrated a self-powered OECT for point-of-care sensing applications. Our work highlights for the first time, a highly efficient, ambient stable solar cell powered OECT, whose transconductance is nearly unaffected by the incident light intensity due to the balanced electron-hole transport in the solar cell. These findings pave the way for higher performance bioelectronics and flexible/wearable electronics.

**INVITED LECTURE**

# In-fiber nanostructures produced by fluid instability

Daosheng Deng
Fudan University

Nanostructures are essential for sophisticated functional devices in fibers and textiles. This talk will present the attainment of various nanostructures during thermal drawing process and post-drawing process, from the perspective of fluid instabilities. One is the classical Rayleigh-Plateau capillary instability for the nanoparticles, and the other is the dewetting instability for the nanofilaments. These results provide theoretical guidance to design and achieve diverse structures for the fiber applications.

**INVITED LECTURE**

# Investigation of the nonlinear effect in the mid-infrared microstuctured optical fibers

Tonglei Cheng
Northeastem University

Soft glass microstructure optical fibers are fabricated based on soft glass materials su ch as tellurite, fluoride and sulfur, which have excellent transmission performance in the midinfrared. Particularly, tellurite and sulfur microstructure optical fibers also have high nonlinear refractive index, which provides favorable condition for developing mid-infrared fiber sensors and nonlinear fiber devices. This report mainly introduces the highl y efficient second harmonic generated in a low-loss tellurite optical fiber, the multi-order soliton and dispersion wave generated in a four-hole tellurite microstructure optical fiber, and the nonlinear optical effect of modulation instability generated in a double-cladding tellurite fiber.

INVITED LECTURE

# $Fe_3O_4$ nanowire arrays on flexible polypropylene substrates and its potential application

Jiangtao Hu
Shanghai Institute of Applied Physics

The in-situ growth of one-dimensional magnetite ($Fe_3O_4$) nanowire (NW) arrays with deterministic and tunable control over their orientation and morphology on a wide range of flexible and low heat resistance substrates is still a challenge . Herein, a facile method of controlling the orientation of $Fe_3O_4$ NW arrays on a polypropylene (PP) nonwoven fabric surface (PP-g-PAO/$Fe_3O_4$) through simultaneous radiation induced graft polymerization and coprecipitation processes was realized. We demonstrated a control over the orientation and geometric properties of coprecipitated $Fe_3O_4$ NWs via complex iron ion using amidoxime groups, which not only endow the material with high durability but also lead to the oriented growth of $Fe_3O_4$ NWs. The PP-g-PAO/$Fe_3O_4$ composite presented a good saturated magnetization, superparamagnetic property, outstanding service life and recyclability, and exhibited two extraordinary sensing capabilities, i.e., UV and magnetic. After the in-situ formation of silver nanoparticles on the surface of the PP-g-PAO/$Fe_3O_4$ (Ag@$Fe_3O_4$-MS), some special properties were observed such as thermal induced switchable conductor-insulator and moisture detection. In detail, at room temperature, the fabricated Ag@$Fe_3O_4$-MS sensor is electrically insulating, while it switches to an electrical conductor when exposed to flames. This strategy can be viewed as the first example of in situ coordination induced growth of $Fe_3O_4$ NW arrays on the surface of a flexible substrate under mild conditions, and nanosensors based on PP-g-PAO/$Fe_3O_4$ hold tremendous prospects for multiparametric sensing platforms.

INVITED LECTURE

# Recent advances in low-loss splicing photonic crystal fibers

Limin Xiao
Fudan University

We demonstrate a novel approach to form ultralow loss and high-strength joints between large mode area solid or hollow-core PCFs and conventional fibers using a modified reverse fiber tapering technique

**INVITED LECTURE**

# All-solid fluorotellurite fibers and their applications

Guanshi Qin
Jilin University

We demonstrated 20-W-level mid-infrared supercontinuum laser source, ultrabroadband supercontinuum generation from 600 to 5400 nm, and broadband amplification in newly-developed all-solid fluorotellurite fibers.

**POSTER PRESENTATION**

# Dielectric properties of polyaniline/silver nanowire composites

Yan An, Jiahao Xin, Xiaofeng Li, Xinfeng Wu, Runhua Fan
Shanghai Maritime University

In this work, polyaniline-coated silver nanowires (Ag@PANI) were prepared by in-situ polymerization of aniline with silver nanowires suspended in it. Then the Ag@PANI/PANI composite was obtained by further mixing of Ag@PANI with PANI. As a comparision, the silver nanowires were mixed with PANI directly to get the Ag/PANI composite. From the SEM observation， we can know that silver nanowires are about 500nm in length and about 20nm in diameter.

The AC conductivity and dielectric constant of Ag@PANI/PANI composites with different mass fraction were measured in the frequency range from 10 kHz to 1 MHz using the LCR meter. Results showed that the electrical conductivity of the composite is proportional to the silver content, while the real part of dielectric constant of the composite is inversely proportional to silver content in the range of 0wt% ~ 15wt%. The DC resistivity of pure PANI, Ag/PANI, Ag@PANI/PANI under illumination were measured by a four-probe tester. The resistivity of the samples could reduced under light condition among which the Ag@PANI/PANI composite has the most obvious change. Its resistivity is reduced by 4.3%.

POSTER PRESENTATION

# The synthesis of electrochromic polyaniline film with fast response and high optical contrast

Xiangyu Huang[1,2], Shaofei Wang[1,2], Suna Fan[1,2], Yaopeng Zhang[1,2]
1.State Key Laboratory for Modification of Chemical Fibers and Polymer Materials
2.College of Materials Science and Engineering， Donghua University

Polyaniline (PANI) has been widely used in electrochromic areas because of its facile synthesis, controllable conductivity, rich color changes, and unique protonation doping mechanism compared to other conducting polymers. The nanostructures with different morphologies can be obtained simply by doping different acids and controlling synthesis conditions in the electrochemical deposition system. Compared with the PANI doped with sulfuric acid, the PANI doped with phytic acid showed a 24% increase in maximum contrast ratios with fast switching time and multicolor diversity. The simple and controllable synthesis method can be extended to further research on electrochromic PANI.

**POSTER PRESENTATION**

# Synthesis and characteristics of red, green, blue fluorescence carbon dots

Ling Jin, Kang Wei
Anhui university of technology

In the present work, novel carbon dots (CDs) were synthesized from precursor comprising L-tyrosine and o-phenylendiamine via a green hydrothermal method. These CDs possess strong orange-red emission, and excellent photostability. Furthermore, the resulting CDs were dispersed in polyvinyl alcohol (PVA) matrix to yield solid state films, where the self-quenching effect in solid state was effectively avoided.

INVITED LECTURE

# 上海交大构 建材料创新人才培养体系的探索与 实践

朱申敏, 杨文红, 张婳
上海交通大学

多年来，上海交大材料学院 紧紧围绕“立德树人”这个根本，以服务“一带一路”战略为契机，始终 贯彻一个理念，坚持双轮驱动，注重三个结合，推进四个协同，构建五大支撑，推动六大转变，有效提高了人才培育质量和国际化水平。强化“四位一体”的培养理念，强调对学生“价值引领、知识探究、能力建设、人格养成”的有机统一；坚持“服务国家需求、赶超世界一流”双轮驱动：注重 与实践教学相结合、与现代信息技术相结合、与创新创业能力培养相结合；推进 科教协同、校企协同、学科协同、中外协同；构建卓越的 师资队伍、课程体系、实践平台、科研项目、国际交流五大人才培养支撑平台。推动了人才培养的 体系、模式、动力、对象、资源、评价的六大转变：人才 培养体系由立足本土向扎根中国融通中外的 国际化 转变；人才培养模式由传统工科模式向更加注重数理基础、拓展交叉复合的“新工科”培养模式 转变；人才培养动力由完成任务式培养转变为“服务国家需求、赶超世界一流”双轮驱动；人才培养对象由局限国内为主转变为 国内国际兼顾；人才培养资源由依托国内校内资源转变为充分整合利用 国际国内校内校外资源；人才培养评价由学校单一评价转变为教师、学生、社会等参与的 多元评价。

INVITED LECTURE

# 基于“五个结合”的地方高校材料类专业拔尖创新人才培养模式的改革与实践

江学良
武汉工程大学

江学良教授2004年7月毕业于上海交通大学获材料学博士学位。同年进入上海交通大学材料学院金属基复合材料国家重点实验室工作。2008年1月调入武汉工程大学材料学院工作，校优秀中青年骨干、2012年科研先进工作者。英国化工协会等。主要从事功能性聚合物、聚合物界面设计以及有机一无机纳米复合材料等方面的研究工作。科研项目：国家自科基金项目?“原位共聚合成聚合物-稀土纳米复合空心球及其改性橡胶制备低频吸声材料研究”，三项省级项目。申请专利6项，授权6项，其中转让一项。已在国内外重点学术期刊上发表论文58篇，其中SCI:16篇，EI:14篇，国内外会议文章12篇。所参与研究的成果获2012年湖北省科技进步二等奖和2012年度中国石油和化学工业联合会科技进步二等奖。2012年教学成果三等奖。担任J. Applied Polymer Science、Polymer International审稿人，参与编写“十一五”材料类国家规划教材1部，培养研究生15人，其中毕业9人，校优秀硕士论文3篇。

INVITED LECTURE

# 以专业认证为抓手，聚焦专业内涵建设

王燕萍
东华大学

东华大学高分子材料与工程专业始终坚持立德树人的根本任务，以专业认证作为抓手，建立完备的人才培养质量三级保障体系，探索发展多维度全过程材料人才培养新模式。专业以培养符合国家战略和产业发展需求的高水平材料类本科专业人才为目标，内化工程认证的质量意识和质量文化，从理念、制度、实施等方面持续改进创新人才培养模式，实施五大措施：

1. 产出导向，健全质量体系保障专业持续发展。建立一系列基于 OBE 理念的制度，积极推进专业培养目标、毕业要求和课程质量的内外部评价，完善课程质量监控和保障体系，从制度上保证专业建设始终坚持以学生为中心，全面提升学生解决复杂工程问题能力，基于社会、学校、院系多层次评价反馈，驱动专业持续改进，提升人才培养质量。
2. 需求牵引，瞄准国家战略和行业发展需求设计课程体系。对接航空航天、新能源等国家战略性新兴产业需求，专业开设多门工程实践课程、前瞻性课程和拓展性课程。采用专题案例教学，全面训练学生工程知识和能力，强化创新意识。通过聘请学科著名学者、行业内知名企业家为学生授课，促进学生知识、能力及综合素养的全面提升。
3. 研教相长，提升学生分析和解决行业前沿复杂工程问题能力。将重大科研成果凝练后引入理论和实践教学，使学生“零距离”接触学科前沿及产业发展新技术；为全体本科生配备科研导师，以课题为抓手培养学生创新能力和解决专业复杂工程问题能力。
4. 校企融合，集聚专业和社会优质资源完善校企协同育人。整合国家重点实验室和国家重点学科优质平台资源用于本科教学。借助“卓越工程师”项目建设，与业龙头企业联合建立“工程实践教育中心”，创建校企协同育人模式，提升学生工程实践能力和职业素养。
5. 互联互动，创建线上线下一体化教学新模式跨时空育人。利用数字化手段，自主开发“互联网+实验教学平台”、“互联网+实习导航系统”、“3D 虚拟仿真系统”等，虚实结合，互联共享，多维度拓展学习时间和空间，引导学生主动学习、终身学习。

专业将继续坚持立德树人的根本任务，紧紧把握新一轮科技革命与产业变革的历史机遇，通过集聚高校、企业、社会各方力量，聚焦专业内涵建设。专业建设将聚焦提升教育质量文化内涵，以“六个一流”为目标，即树立培养一流人才观念，创新一流人才培养模式和一流教育管理体制，打造一流师资队伍和一流工程实践平台，建成国际知名的一流专业，以发展高质量的工程教育和培育高质量工程人才服务新经济发展和国家战略。

INVITED LECTURE

# 智慧平台助力金课建设的思考与实践案例

王婧
化工出版社

本报告共四个部分，一是对金课内涵的思考，从教育部视角和学生视角分别探寻金课的内涵；二是智慧平台的应用，主要介绍易课堂（教育资源平台）和易测（在线题库）两大平台的功能与应用；三是对金课教材的探索，两本教材为案例具体分析金课教材的编写和呈现方式；四材料专业教材建设，简要介绍化学工业出版社材料专业教材建设情况。

INVITED LECTURE

# 团队---科研育人的沃土

魏丽乔
太原理工大学

科学研究、教书育人的系统工程，离不开团队的协同合作。团队中，人才的培养，注重成员不可思议的潜力，让个人的能力得到极大提升，使潜力逐步升华；科学研究方面，注重探索最新前沿科技发展与动态，开拓思维，勤于钻研，集思广益，激发成员最强的科研潜能，促进研究成果的先进性，倡导时效新、效益优。许多成功的事例表明，科研育人因团队合作硕果累累，团队建设因人才培养与成果取得日新月异。

KEYNOTE SPEECH

# 浅谈当前大学教育的若干热点问题

郑强
浙江大学

教育部科学技术委员会委员、教育部高等学校创业教育指导委员会副主任委员、教育部高分子材料与工程教学指导委员会副主任委员、中国化学会—中国力学会流变学委员会副主任委员、中国高分子科学委员会聚合物表征专业委员会副主任委员、教育部全国工程教育专业认证专家、国家自然科学基金委员会工程与材料科学部专家评审组成员、中国材料研究学会理事、中国化学会常务理事等。

《Journal of the Society of Rheology,Japan》中国地区编委，《科学通报》特约编辑，《高分子学报》、《高分子通报》、《材料科学与工程学报》、《功能材料》、《功能高分子学报》、《应用化学》、《塑料工业》、《有机硅材料》、《Journal of Zhejiang University，Science》等期刊编委。

INVITED LECTURE

# 新型聚酯纤维材料的开发及应用

李鑫, 邱志成
中国纺织科学研究院有限公司

聚酯纤维材料是最重要的基础原料之一，它包括聚酯树脂及其纤维、薄膜、片材、工程塑料等。由于聚酯纤维材料机械性能优良、应用广泛，且已形成较成熟的回收体系，随着原材料 PX 和 PTA 产能持续增长，全球聚酯纤维材料产能的年均增长率在未来很长一段时间内将持续保持在 5%以上。

高品质、高功能、环境友好和低成本是聚酯纤维材料的发展方向。因此，聚酯纤维材料产业化关键技术包括：（1）分子设计及其连续聚合技术、绿色高效催化剂，（2）大容量装置柔性化技术，（3）纺程及纤维结构与形态的精确控制技术。

本报告重点介绍了一些已工业化或正在工业化的先进聚酯纤维材料，包括：（1）阻燃纤维、热粘合纤维和导电聚酯纤维等功能性纤维，（2）工业丝、瓶片，（3）原液着色聚酯纤维、再生纤维材料、生物可降解 PBT 共聚酯等环境友好聚酯纤维材料。

最终交流类型：主题报告

**INVITED LECTURE**

# 高性能纤维及复合材料研究与应用进展

余木火
东华大学

东华大学材料学院教授、"973"首席科学家、纤维材料改性国家重点实验室副主任、东华大学研究院副院长，东华大学轻质结构复合材料研究所常务副所长。兼任中国复合材料学会理事、中国复合材料学会科技咨询工作委员会主任、中国复合材料学会增强体专业委员会秘书长。

长期从事高性能纤维及复合材料科研教学工作。曾主持教育部创新团队、NSFC重大研究计划重点项目、863课题、上海市知识服务平台建设等各级课题22项，发表论文207篇，参编著作1部，申报发明专利104项，获上海市科技进步一等奖1项。

INVITED LECTURE

# 海洋生物多糖纤维制备应用研究及产业化进展

夏延致, 纪全, 田星, 全凤玉, 王兵兵, 薛志欣
青岛大学

海藻资源是人类可利用的重要的可再生资源，向海洋要资源，发展蓝色经济，是在土地资源日益紧缺、石油等一次性化石资源日益枯竭、自然环境遭到严重破坏情况下，继续推进我国经济可持续社会发展的重要选择。海藻纤维是以从海带、马尾藻等褐藻类植物中提取的海藻酸盐为原料制备的一种绿色环保的多功能海洋生物多糖纤维新材料，因其丰富来源和优异性能在纺织和医疗领域受到广泛关注。近年来，研究团队系统开展了海藻纤维的纺丝成形、结构与性能、纺织加工及在多领域的应用研究。揭示了海藻酸盐本质阻燃机理，提出金属离子阻燃新理论，发明了金属离子阻燃多糖材料（纤维素纤维及织物、木材、纸、粘合剂等），为高分子材料阻燃化提供了新途径；发明了海藻酸盐分子交联技术，解决了海藻纤维易溶于盐水和洗涤剂的难题；研究海藻酸盐溶液特性与凝胶行为，并应用于海藻多糖的纺丝成形，突破了海藻纤维生产关键技术，研发了成套专用装备，建成了海藻纤维产业化生产线，制定海藻纤维相关技术标准 5 项；产业化项目获得 2016 年山东省技术发明一等奖和中国纺织工业联合会科技进步一等奖。。

INVITED LECTURE

# 无融缩抗氧化聚苯硫醚砜纤维的制备及其性能研究

王明稳
天津工业大学

INVITED LECTURE

# Research and application progress of high quality dope dyed and functional masterbatch technology

Peng Ji[1], Chaosheng Wang[1], Huaping Wang[1], Yiming Xu[2], Wei Xie[2]
1.Donghua University
2.Poly Plastic Masterbatch(Suzhou) Co., Ltd.

China is the world's largest chemical fiber producer, with a production of more than 50 million tons in 2018, accounting for about 70% of the world's total production. With the increasing attention of the state to energy conservation and environmental protection, the problems of high energy consumption and high pollution in traditional dyeing of chemical fiber have become a major adverse factor affecting the development of chemical fiber and downstream textile industry. The dope dyed masterbatch technology can omit most of the dyeing and finishing processes, reduce waste gas and wastewater discharge, greatly save production water and electricity. The dope dyed masterbatch technology conforms to the concept of energy conservation and environmental protection. Because of its remarkable advantages in energy conservation and environmental protection, it is recognized as three kinds of "green fiber" with recycling chemical fiber and biological fiber. In addition, adding functional additives in the fiber dope dyed masterbatch can make the fibers have various properties, such as flame retardant, light blocking, anti-aging, antistatic, breathable and so on, so as to meet the needs of downstream customers for fiber differentiation and functionalization.

INVITED LECTURE

# 锦纶 6 产业链发展及展望

王松林
恒逸石化股份有限公司

本次汇报以己内酰胺锦纶产业链发展及展望为主线，基于市场需求大幅增长的情况下，未来技术发展的大趋势，在己内酰胺产品，主要阐述先进环己酮制造技术、双氧水生产工艺、氨肟化技术。在产业领域方面，支撑先进制造技术，讨论锦纶新聚合工艺、纺丝工艺的发展方向；最终形成从单体、聚合、纤维、应用的全产业链体系。最后对锦纶行业的发展进行了展望，对锦纶制备技术的发展方向、发展模式、制度创新等方面提供建议。

INVITED LECTURE

# 聚乳酸纤维（乳丝）的生产加工、应用及展望
# Production 、Application and Prospect of PLA Fiber

Jie Ren
Tongji University

The dependence of over 60% of petroleum import in China, the shortage of resources and "white pollution" caused by over 60 million tons of chemical fiber consumption, all contribute to the rapid development of new biobased fiber materials to replace traditional petroleum based polymer chemical fiber materials.

Rusi is a new kind of biodegradable green bio-based fiber with high strength, good biocompatibility and biodegradability. The shape of Rusi mainly includes filament and staple. The traditional melt spinning can be used to process Rusi. In addition to good physical and mechanical properties,   and silky breathability of the breast filaments have been widely used in the use of fabrics and nonwovens.

The unique structure of Rusi makes it have the same performance as cotton blended and terylene. Rusi has excellent guide wet again, not sticky to the skin, Rusi underwear helps moisture transfer, not only when in contact with the skin is dry, and can give good shape stability and wrinkle resistance, it is based on the human body contains some lactic acid as a raw material synthesis of lactic acid polymer, not irritate the skin and is beneficial to human body health, very suitable for the raw material of underwear and home textiles.

In addition, Rusi has good elasticity, good shape retention, drape and dyeing properties. The garment fabric made of Rusi spinning or blended with wool fiber has a strong sense of wool and good wrinkle resistance. At the same time, due to the moderate initial modulus of Rusi, the fabric has good drape and feel. Therefore, Rusi is an ideal material for developing garment fabric.

Using dry method, spunbonded method and melt-spray method to form a net, water, acupuncture or thermal bonding, such as reinforcement, can be made into a variety of non-woven products. Because the polylactic acid has a lower melting point, melting point range is very wide in different Rusi (120-170 ℃), and has a good bonding effect, suits made of composite fiber, and application in nonwoven fabric, such as sanitary napkin, pads, diapers, adult incontinence supplies, medical gauze, bandages, medical bed sheets, and other products, not only to be able to get a good bacteriostatic requirements of disposable medical and health products, and its degradation characteristics and can solve the problem of white pollution from disposable goods.

Rusi will gradually replace the traditional petroleum-based chemical fiber and occupy a place of bio-based fiber market in the future.

INVITED LECTURE

# 纤维素纤维的功能化与产业化

李发学
东华大学

纤维素纤维是纺织工业重要的基础原料，具有吸湿性好、亲肤性好、染色鲜艳、环境友好等特性，深受消费者青睐。随着社会需求日益广泛，纤维素纤维的功能化日益受到人们的重视。本文首先介绍了纤维素纤维的国内外发展现状，然后重点概述了纤维素纤维在阻燃、抗菌、超亲水等领域的研究现状、存在技术难题及产业化态势与发展前景等，期望能为我国纤维素纤维的发展带来一些启示。

INVITED LECTURE

# 生物可降解共聚酯的产业化进展

李乃祥, 潘小虎, 戴钧明
中国石化仪征化纤有限责任公司

PBST（聚丁二酸/对苯二甲酸丁二醇酯）、PBAT（聚己二酸/对苯二甲酸丁二醇酯）是一类脂肪族芳香族共聚酯，既有较好的延展性、优良的耐热性和抗冲击性能，同时满足一定丁二酸、己二酸含量要求的共聚酯又具有良好的生物可降解性，是生物可降解聚合物的主要品种。本文对 PBST、PBAT 的性能、应用进行了分析，并介绍了 PBST、PBAT 的制备技术和产业发展现状。

INVITED LECTURE

# 金发先进聚酰胺材料在汽车上的应用

Nanbiao Ye
上海金发科技发展有限公司

This paper mainly introduces the application of PA and HTPA in automobile. Through the different glass fiber reinforced nylon composite materials, the application of KINGFA composite materials in different parts of the car can meet the demanding requirements of the car industry.

INVITED LECTURE

# 智能响应性纤维材料的开发及产业化

王栋[1,2], 赵青华[1], 李沐芳[1]
1.武汉纺织大学
2.东华大学

为了提升纺织材料的产品附加值，满足人们日益增长的生活水平需求，纤维材料正朝着差别化、功能化、智能化及绿色环保方向发展。智能响应性纤维材料是指能够对自然环境中的光、热及湿度等进行感知，并作出相应响应的一类材料。我们结合多种环境响应机制，开发了具有光热转换、光致变色、温致变色以及湿度驱动等功能的智能响应性纤维材料，研究并探索了各类纤维的制备工艺、性能及其应用。目前，光热转换聚酯纤维、光致变色聚酯/尼龙纤维已经实现了产业化生产及应用，具有良好的经济及社会效益，为高附加值功能性纺织品的开发提供了方向。

**INVITED LECTURE**

# 纤维微塑料的产生与预防机制

李方
东华大学

微塑料（Microplastics）污染作为一种新兴的海洋污染问题已经成为全球相关学者研究的热点。微塑料污染遍布全球海洋，从赤道到极地，从河流到大洋，存在于自然水体的表层和深海沉积物中。微塑料及其潜在危害效应沿食物链进行传递、累积，进而影响整个生态系统，最终可能对人类健康产生影响。

不同形态的微塑料中，纤维状微塑料占比最高，主要来源于合成纤维制成的纺织品。在合成纤维的整个生命周期中，高聚物合成，经纺丝成型后制备为长丝、短纤，再经纺纱、织造、染整、裁剪、预洗等加工过程制成为各类针织物、机织物、非织造布、绳索等纺织产品，以及在纺织品使用直至最终处置的整个环节中，均有可能成为微塑料污染物排入环境。纺织品洗涤过程中的纤维脱落，被认为是海洋环境中合成纤维最常见的来源，纤维的脱落量不仅与纤维和纺丝（或纱线）的质量及织物结构相关，洗涤温度、洗涤方式及洗涤剂也起到重要影响。此外，聚酰胺、聚烯烃类的纺织品在农业生产中的使用也是微塑料的重要来源之一，如绳索、鱼线、渔网等农业用具的损耗及丢弃等。

纺织全过程产生的陆源微塑料进入环境，对环境及生物体的生态风险及危害。首先塑料缠绕在生物体的躯体上导致生物幼体畸形或是被生物所误食导致产生假饱状态从而无法进食，最终导致肠道堵塞饥饿而死亡。其次微塑料容易被水环境中浮游动物底栖生物误食，在生物体不同器官和组织中富集转移。微塑料影响 POPs 污染物生物积累的机制主要包括（1）塑料本身作为 POPs 载体来源，通过皮肤、鳃等途径进入生物体，生物摄食过程导致 POPs 转移到有机体中；（2）作为不易消化、不可吸收的亲脂性聚合物，摄食微塑料导致增加 POPs 的排泄和减少身体负担。

基于全生命周期理念，控制纤维微塑料的产生主要从以下几个方面努力：（1）纤维的材质与结构。例如，长丝织物的纤维微塑料产生量明显少于短纤维织物；致密、光滑的高质量面料纤维微塑料产生量明显少于低质量的面料；（2）改善衣物家纺的洗涤方式，例如采用滚筒式洗衣机、轻柔洗模式和中性的洗涤剂的方式；（3）在污水处理厂设置特殊工艺，例如砂滤池、精密过滤器、强化絮凝分离等方式减少尾水中的纤维微塑料；（4）提高产业用纺织品的使用寿命，强化表面涂层技术，减少复合材料中短纤维的用量。

INVITED LECTURE

# 废旧聚酯纤维制品资源再生循环技术发展

陈烨, 王华平, 柯福佑
东华大学

我国聚酯年产量达4000万吨，其废旧纤维制品总储量超过1亿吨，但再生纺丝产能仅1000万吨，再生率不足10%；不仅资源浪费大，且环境负担重，是纺织循环经济发展的重点领域。国际废旧聚酯纤维再生利用主要是实现资源化处理，解决污染问题，重点发展分拣清洗技术及旧衣回用体系；美国日本等开发的以解聚提纯再聚合的化学法技术，由于工艺复杂成本高，产业化推广受限，国内大多采用简单熔融再生纺丝工艺，产品品质低，应用受限，亟需开发兼顾品质与成本的废旧聚酯纤维再生循环产业化方案。